BASIC PROGRAMS
for Scientists
and
Engineers

BASIC PROGRAMS
for Scientists
and
Engineers

Alan R. Miller

Professor of Metallurgy
New Mexico Institute of Mining
and Technology

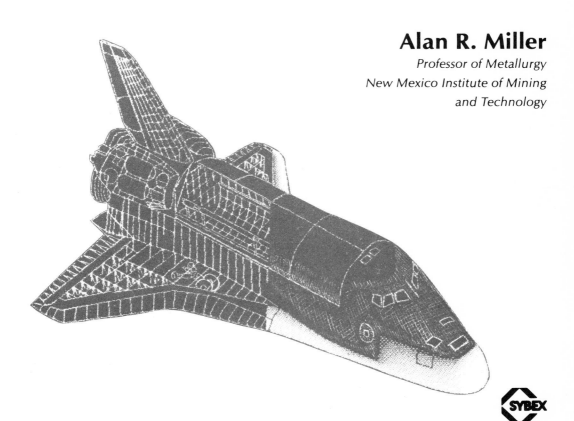

SYBEX

Berkeley • Paris • Düsseldorf

CREDITS

Cover Design by Daniel Le Noury
Technical illustrations by J. Trujillo Smith, Jeanne E. Tennant

NOTICES

Apple is a registered trademark of Apple Computer Corporation.
BASIC-80, Microsoft BASIC, and BASCOM are registered trademarks of Microsoft Consumer Products.
CBASIC is a registered trademark of Compiler Systems, Inc.
DEC-20 is a registered trademark of Digital Equipment Corporation.
Lifeboat 2.2 is a registered trademark of Lifeboat Associates.
North Star BASIC is a registered trademark of North Star Computers, Inc.
PET is a registered trademark of Commodore Business Machines.
TRS-80 is a registered trademark of Tandy Corporation.
Xitan BASIC is a registered trademark of Xitan Systems Ltd.
WordStar is a registered trademark of MicroPro International Corporation.
Z80 is a registered trademark of Zilog, Inc.

SYBEX is not affiliated with any manufacturer.

Library of Congress Card Number: 81-84003
ISBN 089588-073-3
First Edition 1981
Printed in the United States of America
10 9 8 7 6 5 4 3 2 1

Contents

4   *Simultaneous Solution of Linear Equations*   *53*

5   *Development of a Curve-Fitting Program*   *115*

6 *Sorting* 143

7 *General Least-Squares Curve Fitting* 163

8 *Solution of Equations by Newton's Method* 203

9   *Numerical Integration*   *225*

10   *Nonlinear Curve-Fitting Equations*   *255*

11 Advanced Applications:
The Normal Curve, the Gaussian Error Function, The Gamma Function, and the Bessel Function   *277*

Appendix A: Reserved Words and Functions   *305*

Appendix B: Summary of BASIC   *307*

Bibliography   *315*

Index   *317*

Preface

The ideas and material for this book have been developed during my experience teaching sophomore, junior, and senior engineering students over the past 14 years. I have used FORTRAN, BASIC, and Pascal as the computer languages for these courses.

All of the BASIC programs in this book were developed on a Z80 microcomputer. The operating system was the Lifeboat 2.2 version of CP/M. The source programs were developed and executed with Microsoft BASIC-80, Version 5. MicroPro's WordStar was frequently used to modify the ASCII form of the source programs. Most of the programs were also run on other BASICs. These included Microsoft's BASCOM (compiling BASIC), CBASIC, Xitan BASIC, North Star BASIC, and BASIC Plus-4 running on a DEC-20.

The manuscript was created and edited with MicroPro's WordStar running on the same Z80 computer. The BASIC source programs have been incorporated directly into the manuscript from the original source files. Computer printouts shown in the figures were also incorporated magnetically into the manuscript. This was accomplished by altering the CP/M operating system so that console output was written into a block of memory. This block was then saved as a disk file. The final manuscript was submitted to SYBEX in a magnetic form compatible with the photocomposer. Consequently, the manuscript and the BASIC source programs have not been retyped.

I am sincerely grateful for the helpful guidance and suggestions of Rudolph Langer and Douglas Hergert during the development of the manuscript. I would also like to thank Ashok Singh for checking the manuscript, especially the mathematical expressions.

Alan R. Miller
Socorro, New Mexico
August 1981

Introduction

The purpose of this book is twofold: to help the reader develop a proficiency in the use of BASIC, and to provide a library of programs useful for solving problems frequently encountered in science and engineering.

The programs in this book, the second in the SYBEX *Programs for Scientists and Engineers* series, are most valuable to the practicing scientist or engineer. The material is also suitable for a junior- or senior-level engineering course in numerical methods. The reader should have a working knowledge of an applications language such as BASIC, FORTRAN or Pascal. Experience with vector operations and differential and integral calculus will also be helpful.

BASIC is currently the most widely available high-level computer language, especially on microcomputers. Because it is usually implemented as an interpreter, it is a particularly easy language to use: program execution can be interrupted at any point, current values can be printed, and then execution can be resumed.

The lowest level of BASIC is extremely limited. Variable names should be no longer than two characters, and looping control is only available with the **FOR** /**NEXT** and **GOTO** constructions. Fortunately, however, powerful BASICs are becoming more common. These contain features such as long variable names, the **WHILE/WEND** construction and integer typing of variables. The programs in this book were initially developed using a powerful set of BASIC features, and then simplified to the lowest common denominator. Thus, they can be run on all of the common BASICs. In fact, these programs will run on microcomputers such as the Apple, the TRS-80, and the PET, on Microsoft BASIC and on CBASIC. In addition, they will run as well on larger computers such as the DEC-20 with only minor changes to features such as the random number generator and the multiple statement separator. Users are urged to upgrade the programs to the highest possible level. To help with this conversion, suggested alternate names are given in the source programs, using **REM** lines such as the following:

11 **REM** identifiers
12
13 **REM** N1%   NROW%   number of rows
14 **REM** N2%   NCOL%   number of columns
15 **REM** end of identifiers

In lines 13 and 14, the two-character variable names used in the program are followed by longer and more descriptive names suitable for less restrictive BASICs, and then by comments identifying the variables.

The line numbers for the programs in this book have been integrated so that there is a minimum of conflict. It should thus be possible to combine any of the different subroutines into a single program. Duplicate line numbers are only utilized for the same kinds of tasks. The assigned blocks of line numbers are detailed below:

1 - 499	Main program
500 - 880	Input routine
800 - 880	Set up matrix
1000 - 1370	Output routine
2000 - 2410	Numerical integration
2500 - 2810	The error function
3000 - 3550	Sorting routine
3700 - 3790	Conversion to upper case
4000 - 4310	Conversion to square matrix
4400 - 4930	Bessel function of the second kind
5000 - 6150	Matrix solution
7000 - 7990	Plot routine
8000 - 8160	Newton's method
8400 - 8430	Function evaluation
8500 - 8780	Bessel function of the first kind
9000 - 9100	Mean and standard deviation
9200 - 9400	Gamma function
9500 - 9580	Gaussian random numbers
9800 - 9840	Normally distributed random numbers
9999	**END** statement

The reader who is primarily interested in the BASIC programs developed in this book will have no trouble locating them; the sections that contain programs or subroutines are clearly labeled. It should be noted, however, that this book is designed to be read from beginning to end. Each chapter discusses and develops tools that will be used again in subsequent chapters. The mathematical algorithms of each program are methodically described before the program itself is implemented, and sample output is supplied for most of the programs. The following brief descriptions summarize the contents of each chapter.

Chapter 1, **Evaluation of a BASIC Interpreter or Compiler**, identifies weak points in several commercial BASICs, and supplies programs for testing any BASIC. The results will be used to select various constants and operations in later chapters. Also included in Chapter 1 are discussions of the common "**NEXT** without **FOR**" bug and the "10 **GOSUB** 10" problem.

Chapter 2, **Mean and Standard Deviation**, discusses some simple statistical algorithms and presents a program for implementing them. Routines for generating—and testing—both uniform and Gaussian random numbers are also given.

Chapter 3, **Vector and Matrix Operations**, summarizes the operations of vector and matrix arithmetic, including dot product, cross product, matrix multiplication and matrix inversion. Two important programs are developed—one for carrying out matrix multiplication, and another for calculating determinants.

Chapter 4, **Simultaneous Solution of Linear Equations**, presents programs to carry out the algorithms of Cramer's rule, the Gauss elimination method, the Gauss-Jordan elimination method, and the Gauss-Seidel method—all for solving simultaneous equations. In addition, ill conditioning is studied by observing a program that generates Hilbert matrices, and a program is developed for solving equations with complex coefficients.

Chapter 5, **Development of a Curve-Fitting Program**, is the first of a series of chapters on curve fitting. In a good illustration of top-down program development, a linear least-squares curve-fitting program is written and discussed. The program includes routines to simulate data, plot curves, compute the fitted curve, and supply the correlation coefficient.

Chapter 6, **Sorting**, describes and compares several BASIC sorting routines, including two bubble sorts, a Shell sort and a nonrecursive quick sort. A sort routine is incorporated into the curve-fitting program of Chapter 5 to enable the program to handle real experimental data.

Chapter 7, **General Least-Squares Curve Fitting**, extends the curve-fitting program to general polynomial equations, and finds curve fits for three examples: the equations for heat capacity, vapor pressure, and properties of superheated steam.

Chapter 8, **Solution of Equations by Newton's Method**, presents a series of programs that use Newton's algorithm for finding the roots of an equation. This tool will be used again in Chapter 10 for nonlinear curve fitting.

Chapter 9, **Numerical Integration**, develops programs for three different integration methods—the trapezoidal rule, Simpson's rule, and the Romberg method. End correction is also discussed. Simpson's rule will be used in Chapter 11 for evaluating the Gaussian error function.

Chapter 10, **Nonlinear Curve-Fitting Equations**, discusses curve-fitting algorithms for the rational function and the exponential function. Two examples are given—the Clausing factor, and the diffusion equation.

Chapter 11, **Advanced Applications: The Normal Curve, the Gaussian Error Function, the Gamma Function, and the Bessel Functions**, addresses several advanced topics in programming for mathematical applications. This last chapter summarizes and expands upon a number of the concepts presented earlier in the book.

Each chapter also contains exercises designed to extend the reader's comprehension of the material.

For readers who are approaching BASIC for the first time, a summary of the syntax, standard functions, and reserved words of BASIC is included in the appendices. The real educational experience of this book, however, will be gained by carefully working through the programs themselves.

A Note on Typography

Many readers of programming books believe that typeset programs are more likely to contain errors than photographed programs, that errors may be introduced into the programs during the typesetting process. At SYBEX, we have developed a method of typesetting programs (thus enhancing their legibility) without introducing errors. The author submits tested, executable source code on disks. The programs are then run through a formatting program on our in-house computers. The formatting program establishes conventions for spacing, capitalization, etc., so that the programs will have a uniform appearance, without altering their content. Next, the programs (along with the text) are transmitted electronically to our computerized phototypesetter. At no point are the programs actually retyped, so errors that might have been introduced by retyping have been avoided.

The text of this book has been set in the typeface known as Oracle, the programs are in Futura, and the output has been photographed from the actual line-printer output supplied by the author. Reserved words appear in **boldface**. Mathematical expressions (variables, letter constants) appear in *italics*, except for vectors, which appear in boldface roman. For example:

$$A + Bx + Cx^2 = 0$$

$$\mathbf{v} = [1\ 2\ 3]$$

Throughout the book an effort has been made to distinguish typographically between mathematical values and program structures. Thus, juxtaposed in a single paragraph, the reader may see references to the variable x and the BASIC variable X; the vector $\mathbf{v}$ and the BASIC array V; the matrix element v_{ij} and the BASIC array element V(I,J).

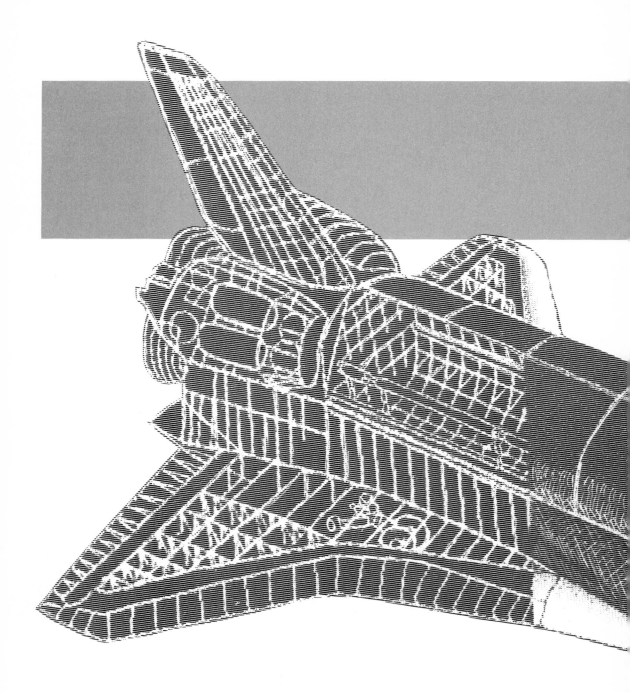

CHAPTER 1

Evaluation of a BASIC
Interpreter or Compiler

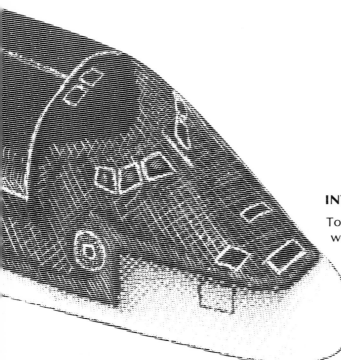

INTRODUCTION

To understand the results of a BASIC program we must be familiar with the limitations of the interpreter or compiler we are using. This is particularly true with scientific application programs such as the ones given in this book. In this first chapter, then, we will present some tools for evaluating the precision and range of BASIC. In fact, the examples given here were derived from several popular BASICs.

PRECISION AND RANGE OF FLOATING-POINT OPERATIONS

Many of the programs in this work are sensitive to the *precision* and *dynamic range* of the BASIC floating-point operations. For example, in one program an algorithm is terminated when a particular term is smaller than a relative tolerance. The formula in this case is:

TERM < SUM * TOL

where TERM is the value of the new term, SUM is the current total, and TOL is an arbitrarily small number known as the *tolerance*.

It is important that the value chosen for the tolerance be within the accuracy of the floating-point operations. Otherwise, the summation step will never terminate. Suppose, for example, that the floating-point operations are performed to a precision of six significant figures. Then, the tolerance must be set to a value larger than 10^{-6}.

The dynamic range of the exponent is a separate matter. Typical binary, floating-point operations are performed with 32 bits of precision. BCD (Binary Coded Decimal) floating-point packages, on the other hand, will usually retain more significant figures and have a greater dynamic range.

We will now present a BASIC program for testing the precision and dynamic range. We will investigate output from several common BASICs to illustrate both mantissa and exponent accuracy.

BASIC PROGRAM: A TEST OF THE FLOATING-POINT OPERATIONS

The program given in Figure 1.1 can be used to determine the precision and the dynamic range of BASIC. Type up the program and execute it. The initial value of X is obtained by dividing $1.0E-4$ by 3. Then, successively smaller and smaller values of X are calculated and displayed

```
10   N% = 18
20   X = 1.0E−4 / 3
30   FOR I% = 1 TO N%
40      X = X/10
50      PRINT I%, X,
60      X = X/10
70      PRINT X
80   NEXT I%
90   END
```

Figure 1.1: A Test of the Floating − Point Operations

on the console. Each succeeding value is obtained by dividing the previous number by 10. The process continues until 36 values have been printed.

Let us now use this program to test the accuracy and range of three different BASICs.

Three Runs of the Program: A Comparison

The initial mantissa is chosen to be 1/3, a repeating fraction that cannot be precisely represented by a floating-point number. Successive multiplications will show the extent of roundoff error. A 32-bit binary, floating-point number typically utilizes three bytes for the mantissa and one byte for the exponent. This usually produces six or seven significant figures of precision and a dynamic range of 10^{+38} to 10^{-38}. The result might look like the output in Figure 1.2. In this example, the dynamic range goes to 10^{-38} and the accuracy of the mantissa is about six significant figures. Notice that no roundoff error is apparent. When floating-point underflow occurs, a value of zero is substituted for the answer and no error message is given.

1	3.33333E-06	3.33333E-07
2	3.33333E-08	3.33333E-09
3	3.33333E-10	3.33333E-11
4	3.33333E-12	3.33333E-13
5	3.33333E-14	3.33333E-15
6	3.33333E-16	3.33333E-17
7	3.33333E-18	3.33333E-19
8	3.33333E-20	3.33333E-21
9	3.33333E-22	3.33333E-23
10	3.33333E-24	3.33333E-25
11	3.33333E-26	3.33333E-27
12	3.33333E-28	3.33333E-29
13	3.33333E-30	3.33333E-31
14	3.33333E-32	3.33333E-33
15	3.33333E-34	3.33333E-35
16	3.33333E-36	3.33333E-37
17	3.33333E-38	0
18	0	0

Figure 1.2: Precision Test: First BASIC

As another example, consider the output in Figure 1.3, from a different BASIC that also uses 32-bit binary floating-point operations. The results in this case are similar to the previous example, except that roundoff error is evident in the mantissa.

The output from a third BASIC that uses 64-bit, BCD floating-point numbers is shown in Figure 1.4. The mantissa contains twelve digits of precision and the dynamic range is 10^{+63} to 10^{-63}. The value of N% at line 10 in the BASIC program is changed to 31 to explore the full potential of the operations.

```
         1              3.33333E-06    3.33333E-07
         2              3.33333E-08    3.33333E-09
         3              3.33333E-10    3.33333E-11
         4              3.33333E-12    3.33333E-13
         5              3.33333E-14    3.33333E-15
         6              3.33333E-16    3.33333E-17
         7              3.33333E-18    3.33334E-19
         8              3.33334E-20    3.33334E-21
         9              3.33334E-22    3.33334E-23
        10              3.33334E-24    3.33334E-25
        11              3.33334E-26    3.33334E-27
        12              3.33334E-28    3.33334E-29
        13              3.33334E-30    3.33334E-31
        14              3.33334E-32    3.33334E-33
        15              3.33334E-34    3.33334E-35
        16              3.33334E-36    3.33334E-37
        17              3.33334E-38    3.33334E-39
        18              0              0
        19              0              0
```

Figure 1.3: Precision Test: Second BASIC

```
         1          3.33333333333E-06    3.33333333333E-07
         2          3.33333333333E-08    3.33333333333E-09
         3          3.33333333333E-10    3.33333333333E-11
         4          3.33333333333E-12    3.33333333333E-13
         5          3.33333333333E-14    3.33333333333E-15
         6          3.33333333333E-16    3.33333333333E-17
         7          3.33333333333E-18    3.33333333333E-19
         8          3.33333333333E-20    3.33333333333E-21
         9          3.33333333333E-22    3.33333333333E-23
        10          3.33333333333E-24    3.33333333333E-25
        11          3.33333333333E-26    3.33333333333E-27
        12          3.33333333333E-28    3.33333333333E-29
        13          3.33333333333E-30    3.33333333333E-31
        14          3.33333333333E-32    3.33333333333E-33
        15          3.33333333333E-34    3.33333333333E-35
        16          3.33333333333E-36    3.33333333333E-37
        17          3.33333333333E-38    3.33333333333E-39
        18          3.33333333333E-40    3.33333333333E-41
        19          3.33333333333E-42    3.33333333333E-43
        20          3.33333333333E-44    3.33333333333E-45
        21          3.33333333333E-46    3.33333333333E-47
        22          3.33333333333E-48    3.33333333333E-49
        23          3.33333333333E-50    3.33333333333E-51
        24          3.33333333333E-52    3.33333333333E-53
        25         ·3.33333333333E-54    3.33333333333E-55
        26          3.33333333333E-56    3.33333333333E-57
        27          3.33333333333E-58    3.33333333333E-59
        28          3.33333333333E-60    3.33333333333E-61
        29          3.33333333333E-62    3.33333333333E-63
        30          3.33333333333E-64    0
        31          0                    0
```

Figure 1.4: Precision Test: Third BASIC

The preceding simple test of the floating-point operations can be supplemented by much more sophisticated techniques. For example, a good test is obtained from the solution to a set of Hilbert matrices. The method is fully described in Chapter 4. However, it will be instructive to consider some actual results at this time.

The exact solution for one particular example consists of six values of unity:

1 1 1 1 1 1

The result calculated by a particular BASIC may be very different from this, however, because of roundoff error from the floating-point operations. The following results were obtained from actual calculations with three different BASICs:

0.999708	1.007900	0.948404	1.130879	0.858209	1.055069
1.00020	0.99523	1.02799	0.93514	1.06502	0.97640
1.00000	1.00000	1.00000	1.00000	1.00000	1.00000

Clearly, there is a measurable difference in the behavior of these three BASICs. The first example demonstrates roundoff error in excess of 14% and the second example shows errors up to 6%. The last example is best of all. Ironically, the BASIC that performed worst with the Hilbert matrix appeared to do quite well with the simpler initial test.

A rather interesting bug contained in several BASICs can limit the range of the SIN function. We will now investigate this phenomenon.

BASIC SIN FUNCTION

A problem can occur with the SIN function as the argument approaches zero. Under this condition, the function should return the value of the argument. However, several BASICs contain an error. They show significant roundoff error or they substitute a value of zero instead of returning the argument.

In the following section we will present a program for studying this problem and test three BASICs with the program.

BASIC PROGRAM: TESTING THE SIN FUNCTION

The program given in Figure 1.5 can be used to check the SIN function of your BASIC. Type up the program and execute it.

```
10    N% = 36
20    X = 1.0E − 4/3
30    FOR I% = 1 TO N%
40        X = X/10
50        PRINT I%, X,SIN(X)
80    NEXT I%
90    END
```

Figure 1.5: Test for Function SIN

Running the SIN Test: Three Examples

This second program, which is similar to the previous one, tests the built-in SIN function. If your SIN function correctly handles small numbers, meaningful values should be returned over the entire dynamic range of the floating-point operations. In this case, floating-point underflow should occur at the same place as it did for the previous test. For the example of Figure 1.6, the dynamic range of the floating-point operations and the limit of the SIN function are both the same.

In Figures 1.7 and 1.8, the SIN function returns incorrect values as the argument approaches zero.

If you find that the SIN function is incorrectly implemented in your

```
1     3.33333E-06    3.33333E-06
2     3.33333E-07    3.33333E-07
3     3.33333E-08    3.33333E-08
4     3.33333E-09    3.33333E-09
5     3.33333E-10    3.33333E-10
6     3.33333E-11    3.33333E-11
7     3.33333E-12    3.33333E-12
8     3.33333E-13    3.33333E-13
9     3.33333E-14    3.33333E-14
10    3.33333E-15    3.33333E-15
11    3.33333E-16    3.33333E-16
12    3.33333E-17    3.33333E-17
13    3.33333E-18    3.33333E-18
14    3.33334E-19    3.33334E-19
15    3.33334E-20    3.33334E-20
16    3.33334E-21    3.33334E-21
17    3.33334E-22    3.33334E-22
18    3.33334E-23    3.33334E-23
19    3.33334E-24    3.33334E-24
20    3.33334E-25    3.33334E-25
21    3.33334E-26    3.33334E-26
22    3.33334E-27    3.33334E-27
23    3.33334E-28    3.33334E-28
```

Figure 1.6: SIN Test: First BASIC

24	3.33334E-29	3.33334E-29
25	3.33334E-30	3.33334E-30
26	3.33334E-31	3.33334E-31
27	3.33334E-32	3.33334E-32
28	3.33334E-33	3.33334E-33
29	3.33334E-34	3.33334E-34
30	3.33334E-35	3.33334E-35
31	3.33334E-36	3.33334E-36
32	3.33334E-37	3.33334E-37
33	3.33334E-38	3.33334E-38
34	3.33334E-39	3.33334E-39
35	0	0
36	0	0

Figure 1.6: SIN Test: First BASIC (cont.)

1	3.33333E-06	3.37056E-06
2	3.33333E-07	3.74507E-07
3	3.33333E-08	0
4	3.33333E-09	0
• • •		

Figure 1.7: SIN Test: Second BASIC

1	3.33333333333E-06	3.3333333E-06
2	3.33333333333E-07	3.333333E-07
3	3.33333333333E-08	3.33333E-08
4	3.33333333333E-09	3.3333E-09
5	3.33333333333E-10	3.333E-10
6	3.33333333333E-11	3.33E-11
7	3.33333333333E-12	3.3E-12
8	3.33333333333E-13	3E-13
9	3.33333333333E-14	0
10	3.33333333333E-15	0
• • •		

Figure 1.8: SIN Test: Third BASIC

BASIC, you may have to use it with care. The subroutine given in Figure 8.15 shows how this can be done. If you find a problem with your built-in SIN function, you can also expect to find a similar problem with the built-in TAN function.

BASIC PROGRAM: THE "NEXT WITHOUT FOR" BUG

Several BASICs contain a curious error in the operation of **FOR/NEXT** loops. Whenever possible, **FOR** loops should be used in preference to the **GOTO** construction because of the greater execution speed. **FOR** loops are especially suitable when the number of loops is known exactly.

In some applications, however, the required number of loops is not initially known. Looping is terminated when the results of a particular calculation reach a desired result. In this situation, the **WHILE/WEND** construction is the perfect choice. Unfortunately, not all BASICs incorporate such a feature. The **FOR** loop, therefore, is used instead. In this case, however, the upper limit for the loop index is set to an excessively large value. When the desired condition is satisfied, the loop is prematurely terminated by an **IF** statement.

Consider the first loop in the program given in Figure 1.9 (lines 20 to 40). The upper limit of the loop index is set to the value of 10. But the **IF** statement will terminate the loop prematurely when the index reaches a value of 9. A separate pair of nested loops (lines 60 to 100) are executed next. The inner loop index uses the same variable, I%, as the first loop, which was never terminated. The program will operate properly on some BASICs, but on others execution of the second loop will terminate with the error message:

NEXT WITHOUT **FOR**

```
 10   N% = 10
 20   FOR I% = 1 TO N%
 30      IF I% = 9 THEN 50
 40   NEXT I%
 50   PRINT "At line 50, I = "; I%
 60   FOR J% = 1 TO 2
 70      FOR I% = 1 TO N%
 80         PRINT I%;J%,
 90      NEXT I%
100   NEXT J%
110   END
```

Figure 1.9: Looking for the NEXT without FOR Error

If this bug is present in your BASIC, you must be careful when prematurely terminating a **FOR** loop. One solution is to avoid using an unterminated loop variable for another loop. The other solution is to choose a **GOTO** construction in preference to **FOR/NEXT** whenever a loop will be prematurely terminated. The problem is complicated by an apparent inconsistency: if the outer **FOR** loop using the J% index is replaced by a **GOTO** statement, the problem will not occur even though it is the inner loop that uses the I% index.

A GOSUB WITHOUT RETURN

When a **GOSUB** command is executed in BASIC, the program flow jumps to the indicated line number. The return address is typically pushed onto the computer's stack at this time. At the conclusion of the subroutine, a **RETURN** statement pops the calling location's address from the stack. This will effect an orderly return to the original location in the program. Subroutines may be nested within other subroutines. When one subroutine calls another, the current address of the calling subroutine is also pushed onto the stack. This process may be repeated several times. Eventually, a sequence of **RETURN** instructions will be executed so that control will return to the main program. The stack will be popped once for each **RETURN** instruction.

The allowable nesting depth for subroutines is usually limited only by the available stack space. Since BASIC typically places the stack at the top of usable memory, there is normally no practical limitation to the nesting level. Try the following one-line program on your BASIC:

10 **GOSUB** 10

While this is a meaningless statement, you can use it to test the stack management of your BASIC. There will be one of three responses when this program is executed.

Some BASICs will immediately stop execution and report a syntax error. This, of course, is the preferred response. Other BASICs will attempt to execute the instruction. In this case, the stack will be pushed, and line 10 will be executed repeatedly. As the process continues to cycle, the stack will grow downward in memory. Eventually, the stack will grow into the BASIC program. One of two things can happen at this point. Some BASICs will stop execution and report an ''OUT OF MEMORY'' error or a ''MEMORY FULL'' error. There are a few BASICs, however, that will simply die at this point. In this case, your program may be destroyed.

SUMMARY

In this chapter we have written a number of evaluative tools designed for BASIC. We began with a program to test floating-point operations; we ran this program on three different BASICs and compared the results. We then discovered an interesting quirk of the SIN function of some BASICs and we devised a program for testing this function. Finally, we explored the elusive **NEXT** without **FOR** error and the 10 **GOSUB** 10 problem.

EXERCISES

1-1: Write a short routine to test the square root function, SQR. Input a value from the keyboard, take the square root, and print the result. Square this value and print the answer.

1-2: Test the tangent and arctangent functions by writing a program to take the combinations TAN(ATN(X)) and ATN(TAN(X)).

1-3: Read values from the keyboard and take EXP(LOG(X)) and LOG (EXP(X)).

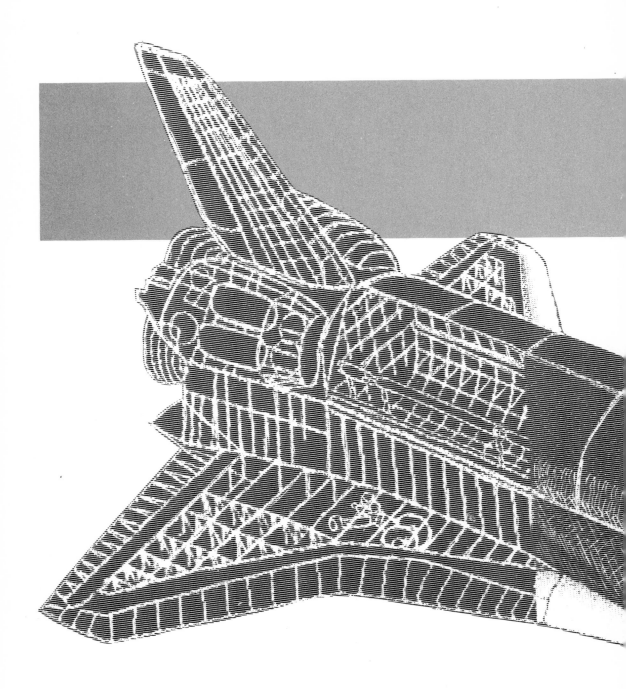

CHAPTER **2**

Mean and Standard Deviation

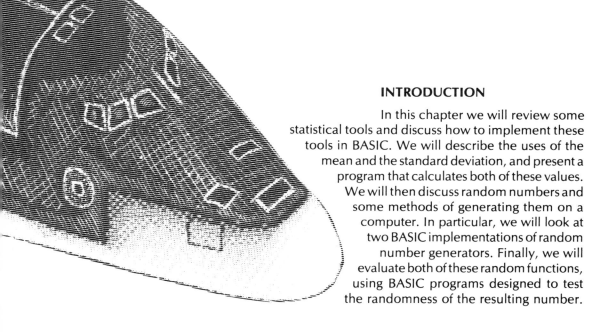

INTRODUCTION

In this chapter we will review some statistical tools and discuss how to implement these tools in BASIC. We will describe the uses of the mean and the standard deviation, and present a program that calculates both of these values. We will then discuss random numbers and some methods of generating them on a computer. In particular, we will look at two BASIC implementations of random number generators. Finally, we will evaluate both of these random functions, using BASIC programs designed to test the randomness of the resulting number.

THE MEAN

We often use a single number, called the *average* or *mean value*, to summarize a particular group of data. The mean is calculated by adding all of the items in the group and then dividing by the number of items. The formula is:

$$\bar{y} = \frac{\sum_{i=1}^{N} y_i}{N} \tag{1}$$

where y_i is the set of data containing N elements. The symbol $\bar{y}$ (pronounced y-bar) is the resulting mean.

On the other hand, when there is a uniform *continuous distribution* of the data, the mean can be determined by integration. Consider the function $y = f(x)$ over the interval from limit a to limit b. The mean value of y is constant over this interval. Consequently, the area under the mean will be equal to the area under the curve $f(x)$:

$$\bar{y}(b - a) = \int_a^b f(x)dx$$

Therefore, the average value of y is

$$\bar{y} = \frac{\int_a^b f(x)dx}{b - a}$$

Dispersion from the Mean

Sometimes all of the data are close to the mean. In other cases, there is a great range of values. As an example of the latter, consider the reporting of weather. The average annual rainfall of San Francisco is said to be 19 inches and the average annual snowfall of New York City is given as 30 inches. But some years are very wet and others are very dry. The average annual temperature of Albuquerque and San Francisco is exactly the same: 57 degrees, but these two cities have very different climates.

As another example, consider a particular brand of breakfast cereal that contains the statement:

Net weight 16 ounces

on the box. Suppose that an inspector from the Bureau of Weights and Measures decides to check this brand of cereal in a grocery store. Several boxes are opened and the contents are weighed. Some boxes are found to contain exactly 16 ounces, but others have 15 ounces or

17 ounces. Should the boxes that contain only 15 ounces be con-
fiscated as examples of short weight? If the contents of 100 boxes of
cereal are weighed, the resulting frequency distribution might look like
the curve in Figure 2.1.

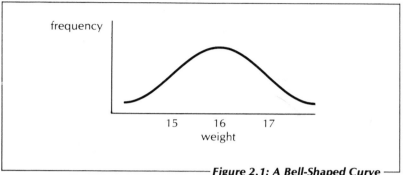

Figure 2.1: A Bell-Shaped Curve

The average weight is 16 ounces, but some of the cereal boxes weigh
more than the mean value and others weigh less. Furthermore, there
are very few boxes that weigh more than 17 ounces or less than 15
ounces.

The frequency distribution shown in Figure 2.1 is *bell shaped*. The
curve shows a *Gaussian* or *normal* distribution. This behavior is typical
of random variation about a mean value. The equation of this bell-
shaped curve is related to the Gamma function and the Gaussian error
function, which we will study in Chapter 11.

Now, suppose that a second type of breakfast cereal is also tested.
The results, this time, might look like the curve in Figure 2.2. The data
again show a frequency distribution that is bell shaped with a mean
value of 16 ounces. But this time, there is a larger dispersion in weights.

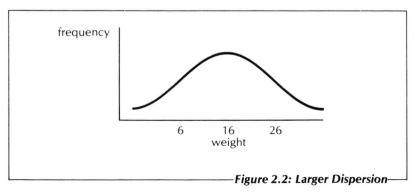

Figure 2.2: Larger Dispersion

Some boxes are as heavy as 26 ounces while others are as light as 6 ounces.

Clearly, there is a difference in the packaging of the first and second types of cereal, even though they both have the same average weight. Something else, besides the mean value, is needed to describe the distribution. We need something that describes the *dispersion*. The tool that we can use, the standard deviation, is described in the next section.

THE STANDARD DEVIATION

The *standard deviation* is a measure of dispersion about the mean value. A large standard deviation means a large dispersion; a small deviation means a small dispersion. The symbol for the standard deviation is the lower-case Greek letter sigma. The standard deviation is defined by the relationship:

$$\sigma = \sqrt{\frac{\sum_{i=1}^{N} (\bar{y} - y_i)^2}{N - 1}} \qquad (2)$$

where $\bar{y}$ is the mean and y_i is the set of N data.

Equation 2 demonstrates the meaning of the standard deviation. The square of the difference between each element and the mean value is important. If the elements are closely grouped about the mean, then this difference will be small. The corresponding standard deviation will also be small. On the other hand, if data are spread far from the mean, this difference will be large. The resulting standard deviation will also be large.

The standard deviation can be used quantitatively to describe the dispersion of a set of data. For example, a range of one standard deviation on either side of the mean includes about 68% of the sample. About 95% of the values lie within two standard deviations of the mean, and almost all of the values, 99.7%, lie within three standard deviations of the mean.

$$\bar{y} \pm \sigma \qquad 68\%$$

$$\bar{y} \pm 2\sigma \qquad 95\%$$

$$\bar{y} \pm 3\sigma \qquad 99.7\%$$

Suppose that we want to select a type of steel for constructing a bridge. Tensile tests are conducted to determine the strength. One steel is found to have a strength of 450 MPa (megapascals), with a standard deviation of 10 MPa. The results indicate that 99.7% of the pieces are expected to have a strength in the range 420 to 480 MPa, that is, within three sigmas. Consequently, a design could be based on a strength of 420 MPa, three sigmas below the mean value.

But suppose that a second type of steel was found to have a mean strength of 500 MPa. Is it a better steel than the first? What if the standard deviation for this second steel is 40 MPa? This larger sigma means a larger spread in values. A strength of three sigmas below the mean of this second steel is only 380 MPa. Thus, this second steel is not as good, even though its mean value is higher.

Metals used for construction typically have small sigmas, but brittle materials, such as concrete and glass, are very different. They usually have large sigmas. Suppose that tests are made on a type of glass suitable for an office door. The average breaking strength is 120 MPa. If the sigma is only 40 Mpa, then three sigmas below the mean gives a value of zero. Thus, it is important to consider the standard deviation as well as the mean value.

In this section we have learned the meaning and the importance of the standard deviation. We will now see how to calculate this value and we will design a BASIC program to perform the calculation.

Calculation of the Standard Deviation

While Equation 2 correctly demonstrates the meaning of the standard deviation, it is an inferior method of calculation. The subtraction of each element from the mean value will cause a loss of significance. This is especially important for points that are very close to the mean. That is, the smaller the value of sigma, the greater the problem will be. Another disadvantage of Equation 2 is that a calculation of the average is required before any subtractions can be performed. This calculation cannot be performed until all of the data are available.

A better method of calculating the standard deviation can be derived by expanding the numerator of Equation 2:

$$\Sigma\, (\bar{y}^2 - 2\bar{y}y_i + y_i^2)$$

Now, by combining the above with Equation 1 we have:

$$\left(\frac{\Sigma y_i}{N}\right)^2 N - 2\left(\frac{\Sigma y_i}{N}\right)\Sigma y_i + \Sigma y_i^2$$

since $\bar{y} = \Sigma y_i/N$. The resulting formula is:

$$\sigma = \sqrt{\frac{\Sigma y^2 - \Sigma y \, \Sigma y/N}{N - 1}} \tag{3}$$

With this method, two running totals are kept: the sum of the individual values and the sum of the squares of the values.

BASIC PROGRAM: MEAN AND STANDARD DEVIATION

The program shown in Figure 2.3 can be used to calculate the mean and standard deviation of a set of numbers. The main program (lines 10 to 100) is designed to perform four main tasks:

1. Set the dimension and the length of the array X (which will hold the values to be averaged) after defining M1%, (the maximum number of values that may be entered):

   20 M1% = 80
   30 **DIM** X(80)

   Some BASICs allow the length specifier (80 in this example) to be expressed as a variable. In this case we can write:

   20 M1% = 80
   30 **DIM** X(M1%)

   The advantage of this feature is that it allows us to change the length of the array simply by changing the value of M1%.

2. Print the title.

3. Call two subroutines, one (at line 500) to ask the user for data, and another (at line 9000) to calculate the mean and standard deviation.

4. Print the results.

The subroutine at line 9000 finds the sum of the values (S1) and the sum of the squares of the values (S2) in one **FOR** loop containing the two lines:

   9050 S3 = S3 + X(I%)
   9060 S4 = S4 + X(I%) * X(I%)

Immediately following this loop, the mean (M2) and standard deviation

(S1) are calculated:

9080 M2 = S3 / N%

9090 S1 = SQR((S4 − S3 ∗ S3 / N%) / (N% − 1))

Line 9080 corresponds to Equation 1; line 9090 corresponds to Equation 3.

The program begins by asking the user for the number of values to be entered. If the answer exceeds the declared length of the array X, then the question is asked again. After the values have been entered, the mean and standard deviation are determined.

If many values are to be entered, an alternate method might be considered. In this case, the computer program could count the number of values as they are entered. A special value, such as a number less than −20000, could be used to signal the end of the list.

```
 10   REM find mean and standard deviation, Apr 25, 1981
 11   REM identifiers
 13   REM       M1%    MAX%       maximum length for X
 14   REM       M2     MEAN       mean of vector X
 15   REM       S1     STD        standard deviation
 16   REM       S3     SUMX       sum of X
 17   REM       S4     SUMSQ      sum of X squared
 18   REM end of identifiers
 19   REM
 20   M1% = 80
 30   DIM X(80)
 40   PRINT "Calculation of mean and standard deviation"
 50   GOSUB 500 : REM input data
 60   GOSUB 9000 : REM find mean and standard deviation
 70   PRINT
 80   PRINT " For "; N%; " points, mean = "; M2;
 90   PRINT " sigma = "; S1
100   GOTO 50
500   REM input the data
510   INPUT " How many points"; N%
515   IF (N% > M1%) THEN 510
520   IF (N% < 2) THEN 9999
```

Figure 2.3: Calculation of the Mean and Standard Deviation

```
530     FOR I% = 1 TO N%
540        PRINT I%;
550        INPUT X(I%)
560     NEXT I%
570     RETURN : REM from input
9000    REM find mean and standard deviation of N points in array X
9001    REM identifiers
9004    REM     M2      MEAN         mean of vector X
9006    REM     S1      STD          standard deviation
9007    REM     S3      SUMX         sum of X
9008    REM     S4      SUMSQ        sum of X squared
9009    REM end of identifiers
9010    REM
9020    S3 = 0
9030    S4 = 0
9040    FOR I% = 1 TO N%
9050        S3 = S3 + X(I%)
9060        S4 = S4 + X(I%) * X(I%)
9070    NEXT I%
9080    M2 = S3 / N%
9090    S1 = SQR((S4 - S3 * S3 / N%) / (N% - 1))
9100    RETURN : REM from mean and standard deviation
9999    END
```

Figure 2.3: Calculation of the Mean and Standard Deviation (cont.)

Type up the program and run it. Input the five numbers 1,2,3,4,5, and verify that the mean is 3 and the standard deviation is 1.58. The results should look like Figure 2.4.

```
Calculation of mean and standard deviation
How many points? 5
1 ? 1
2 ? 2
3 ? 3
4 ? 4
5 ? 5

For  5  points, mean =  3  sigma =  1.58114
```

Figure 2.4: Execution of the Mean and Standard Deviation Program

This completes our introduction to the mean and the standard deviation. We will later return to these subjects; but we will now proceed to the subject of random numbers and random number generators.

RANDOM NUMBERS

A set of random numbers can sometimes be used to test a computer program if actual data are not available. For example, suppose that we have written a program to fit a straight line through a set of experimental data. We could generate a set of points along a straight line. Then, we could selectively move the points off the line by using a random number generator.

BASIC FUNCTION: A RANDOM NUMBER GENERATOR

Random number generators are usually incorporated in BASIC interpreters and compilers. A random number generator will be required for some of the programs given in this book. If your BASIC does not include a random number generator, then you can use the subroutine shown in Figure 2.5 for this purpose. The program has been written as a subroutine rather than a function, since many BASICs do not allow multi-line functions.

```
9800    REM random number generator, initialize seed to 4
9801    REM identifiers
9803    REM      R1      RAN           Random number
9804    REM      S2      SEED          random number seed
9805    REM end of identifiers
9806    REM
9810    R1 = (S2 + 3.14159)^5.04
9820    R1 = R1 - INT(R1)
9830    S2 = R1
9840    RETURN : REM from random number generator
```

Figure 2.5: A BASIC Subroutine for Generating Random Numbers

This random number generator will return a sequence of numbers in the range of zero to unity. Line 9810 uses the following equation:

$$y = (x + 3.14159)^{5.04}$$

For a BASIC that does not have the exponentiation operator, an equivalent expression is:

$$y = e^{5.04 \, (\ln(x \, + \, 3.14159))}$$

In this case the BASIC expression would be:

9810 R1 = EXP(5.04 * LOG(S2 + 3.14159))

Line 9820 produces a number between zero and unity by subtracting the truncated integral part of the original value of R1. This subroutine is called with the statement **GOSUB** 9800 and the desired value is returned in the variable R1. The variable S2 is the seed for generating the next random number in the sequence. It should initially be set to a value of 4. After the first call to the random number generator, the seed is automatically set to the previously returned random number. You may want to experiment with the value of 5.04 on line 9810. A slight change may improve the performance.

Now that we have learned how our random number generator works, let us look at a program that tests the quality of the random number generator.

BASIC PROGRAM:
EVALUATION OF A RANDOM NUMBER GENERATOR

Whether you use the random number generator given in Figure 2.5 or the one supplied with your BASIC, you should perform a simple test to see how reasonable the results are. The mean value of a set of random numbers ranging from zero to unity should, of course, be one-half. Furthermore, the standard deviation should be the reciprocal of the square root of 12, a value of 0.2887.

The program shown in Figure 2.7 can be used to test a random number generator. The subroutine at line 500 calls the random number generator 48 times and stores the returned values in the array X. The main program then calls the subroutine at line 9000 for the calculation of the mean and the standard deviation. Both of the subroutine calls and a print statement are contained in a **FOR** loop of the main program (lines 80 to 120) which repeats the process 20 times.

The argument to function RND in line 520 is shown as unity. However, you may need to change this to a value of zero for some versions of BASIC or remove it entirely for other versions.

Running the Program

Type up the program shown in Figure 2.7. The random number generator given in Figure 2.5 should also be included if necessary. In this case the seed (S2) should be defined near the beginning of the program:

20 S2 = 4

Furthermore, a call to the random number subroutine should be placed at line 515, and line 520 should be changed. These two lines will then look like:

515 **GOSUB** 9800

520 X(I%) = R1

Now test your random number generator. Random number generators are sometimes called pseudo-random number generators to emphasize the fact that they are not truly random. You should not expect to obtain a mean of one-half and a standard deviation of 0.2887 for each grouping of 48 values. The first part of the output from the program might look like Figure 2.6:

```
            mean       std dev
            (0.5)      (0.2887)
            ===================
            0.5305     0.3124
            0.4766     0.3211
            0.3790     0.2736
            0.5005     0.3166
            0.4654     0.2735
            0.5444     0.2987
            0.4699     0.2846
            0.5235     0.3140
            0.5058     0.3072
            0.5075     0.3162
            0.4838     0.2999
```

Figure 2.6: Testing a Random Number Generator

Notice that the mean ranges from 0.38 to 0.54, and the standard deviation has values from 0.27 to 0.32. Other random number generators may be better or worse than this.

```
10   REM find mean and standard deviation, Apr 19, 1981
11   REM identifiers
13   REM      M2      MEAN        mean of vector X
14   REM      S1      STD         standard deviation
15   REM end of identifiers
20   REM
```

Figure 2.7: Program to Test the Random Number Generator

```
30     DIM X(80)
40     N% = 48
50     PRINT "    mean      std  dev"
60     PRINT "     (0.5)        (0.2887)"
70     PRINT "   = = = = = = = = = ="
80     FOR J% = 1 TO 20
90        GOSUB 500 : REM input data
100       GOSUB 9000 : REM find mean and standard deviation
110       PRINT USING "  ##.####    ##.####"; M2, S1
120    NEXT J%
130    GOTO 9999
500    REM input the data
510    FOR I% = 1 TO N%
520       X(I%) = RND(1) : REM may need to be RND(0) or RND
530    NEXT I%
540    RETURN : REM from input
9000   REM find mean and standard deviation of N points in array X
9001   REM identifiers
9003   REM        M2      MEAN        mean of vector X
9004   REM        S1      STD         standard deviation
9005   REM        S3      SUMX        sum of X
9006   REM        S4      SUMSQ       sum of X squared
9007   REM end of identifiers
9010   REM
9020   S3 = 0
9030   S4 = 0
9040   FOR I% = 1 TO N%
9050      S3 = S3 + X(I%)
9060      S4 = S4 + X(I%) * X(I%)
9070   NEXT I%
9080   M2 = S3 / N%
9090   S1 = SQR((S4 - S3 * S3 / N%) / (N% - 1))
9100   RETURN : REM from mean and standard deviation
9800   REM include random number generator if needed
9999   END
```

Figure 2.7: Program to Test the Random Number Generator (cont.)

Using Pi to Produce Random Numbers

A short sequence of random numbers can be obtained from the first fifteen digits of pi:

3.14159265358979

The following mnemonic makes it easy to remember the sequence. The number of letters in each word is equal to the corresponding digit of pi:

YES, I NEED A DRINK, ALCOHOLIC, OF COURSE, AFTER

3 . 1   4   1   5   9   2   6   5

THE HEAVY SESSIONS INVOLVING QUANTUM MECHANICS

3   5   8   9   7   9

This sequence has a mean value of 5.1 and a standard deviation of 2.8.

Our second type of random number generator, described in the next section, is designed to simulate experimental data.

Gaussian Random Numbers

Sampling errors, that is, errors made during measurement, can be of any size. However, small errors are more likely than large errors. For example, suppose that a table with an actual length of six feet is measured with a ruler. An error of one foot is less likely than an error of one inch. An error of ten feet is even less likely. Thus, measured values that are further from the correct value are less likely than values that are closer. Consequently, if we want to simulate experimental data with a random number generator, then the numbers should not be uniformly spaced. In fact, however, the usual random number generator produces a uniform set of numbers over the interval 0 to 1.

What is needed for the simulation of experimental data is a set of random numbers that are grouped about the mean. Thus, the frequency distribution should be bell shaped; it should have a normal or Gaussian distribution. Fortunately, it is fairly easy to produce a Gaussian distribution of random numbers from an ordinary random number generator.

Consider a sequence of 12 random numbers. It is highly unlikely that all 12 will have a value of zero. Similarly, it is very unlikely that they will all be unity. Suppose that we generate a new number from the sum of 12 uniformly distributed random numbers. Since the average value of the original numbers is one-half, the sum of 12 random numbers is likely to be about 12 times one-half, or 6. Thus, Gaussian random numbers can be generated from an ordinary random number generator. We will now look at a BASIC implementation of such a function.

BASIC PROGRAM:
GENERATING AND TESTING GAUSSIAN RANDOM NUMBERS

Gaussian random numbers can be obtained by summing 12 uniformly distributed random numbers and subtracting the value of 6. A set of such numbers should have a mean value of zero and a standard deviation of unity. Other values for the mean and standard deviation can readily be chosen. The formula for calculating each random number, GRND, is:

$$GRND = (GSUM - 6) * DSTD + DMEAN$$

In this expression, GSUM is the sum of 12 uniformly distributed random numbers, DSTD is the desired standard deviation, and DMEAN is the desired mean. If each Gaussian random number is formed from more than 12 numbers, then there is an additional complication. In this case, the formula becomes:

$$GRND = DSTD * (GSUM - NUM /2) * SQR(12 / NUM) + DMEAN$$

where NUM is the number of uniformly distributed random numbers used to obtain each Gaussian random number.

A Gaussian random number generator can be used in conjunction with the regular random number generator to produce a series of Gaussian-distributed random numbers. There are two parameters that must be supplied by the calling program: the desired mean and the standard deviation of the resulting numbers. An alternate approach would be to set these values in the subroutine.

A BASIC program for generating and testing a sequence of Gaussian random numbers is given in Figure 2.8. The main program sets the mean (D1) to 10 and the standard deviation (D2) to 0.5. The loop that controls the subroutine calls (lines 80 to 120) is the same as in the program of Figure 2.7 The subroutine at line 500, however, now calls another subroutine 48 times to produce an array of 48 Gaussian random numbers. This last subroutine, at line 9500, uses a **FOR** loop to call the random number generator, RND, 12 times and then find the sum of the 12 numbers:

```
9540 FOR I2% = 1 TO 12
9550  S5 = S5 + RND(1)
9560 NEXT I2%
```

As before, the argument to RND may need to be set to 0 or removed entirely.

For a BASIC that has no RND function, we would incorporate our random number generator into the program. The **FOR** loop would then become:

```
9540 FOR I2% = 1 TO 12
9545   GOSUB 9800
9550   S5 = S5 + R1
9560 NEXT I2%
```

Either way, line 9570 uses the algorithm we described above to produce a Gaussian random number.

The program can be revised slightly to accept the desired mean and the desired standard deviation as input. Lines 43 and 45 would become:

```
43 INPUT "Desired mean"; D1
45 INPUT "Desired standard deviation"; D2
```

Line 60 would also have to be deleted or changed to print the values entered by the user:

```
60 PRINT " ("; D1; ") ("; D2; ")
```

```
10   REM find mean and standard deviation, Apr 19, 1981
11   REM identifiers
13   REM     D1      DMEAN       desired mean
14   REM     D2      DSTD        desired stand. deviation
15   REM     G1      GRND        Gaussian random number
16   REM     M2      MEAN        mean of vector x
17   REM     S1      STD         standard deviation
18   REM     S3      SUMX        sum of x
19   REM     S4      SUMSQ       sum of x squared
20   REM     S5      SUMG        sum of random numbers
21   REM end of identifiers
22   REM
30   DIM X(80)
```

Figure 2.8:
Generating and Testing Gaussian-Distributed Random Numbers

```
40    N% = 48
43    D1 = 10
45    D2 = .5
47    PRINT '' Gaussian—distributed  random  numbers''
50    PRINT ''      mean        std  dev''
60    PRINT ''       (10)         (0.5)''
70    PRINT ''    = = = = = = = = = =''
80    FOR J% = 1 TO 20
90    GOSUB 500 : REM input data
100   GOSUB 9000 : REM find mean and standard deviation
110   PRINT USING ''   ##.####     ##.####''; M2, S1
120   NEXT J%
130   GOTO 9999 : REM done
500   REM input the data
510   FOR I% = 1 TO N%
515      GOSUB 9500 : REM Gaussian random number
520      X(I%) = G1
530   NEXT I%
540   RETURN : REM from input
9000  REM find mean and standard deviation of N points in array X
9001  REM identifiers
9004  REM      M2      MEAN        mean of vector X
9006  REM      S1      STD         standard deviation
9007  REM      S3      SUMX        sum of X
9008  REM      S4      SUMSQ       sum of X squared
9009  REM end of identifiers
9010  REM
9020  S3 = 0
9030  S4 = 0
9040  FOR I% = 1 TO N%
9050     S3 = S3 + X(I%)
9060     S4 = S4 + X(I%) * X(I%)
9070  NEXT I%
```

Figure 2.8: Generating and Testing Gaussian-Distributed Random Numbers (cont.)

```
9080    M2 = S3 / N%
9090    S1 = SQR((S4 — S3 * S3 / N%) / (N% — 1))
9100    RETURN : REM from mean and standard deviation
9500    REM generate Gaussian random numbers
9510    REM desired mean and standard deviation are D1 and D2
9511    REM number is returned in G1
9512    REM identifiers
9513    REM      D1      DMEAN      desired mean
9514    REM      D2      DSTD       desired stand. deviation
9515    REM      G1      GRND       Gaussian random number
9517    REM      S5      SUMG       sum of random numbers
9518    REM end of identifiers
9520    REM
9530    S5 = 0
9540    FOR I2% = 1 TO 12
9550       S5 = S5 + RND(1) : REM may need to be RND(0) or RND
9560    NEXT I2%
9570    G1 = (S5 — 6) * D2 + D1
9580    RETURN : REM from Gaussian random number generator
9999    END
```

Figure 2.8: Generating and Testing Gaussian-Distributed Random Numbers (cont.)

The area under the normal distribution curve is related to the standard deviation. It can be obtained from the Gaussian error function developed in Chapter 11.

SUMMARY

We began this chapter with a discussion of two important statistical tools, the mean and the standard deviation. Our discussion led us to a BASIC program for calculating these values. Then we discovered how to write two different BASIC random number generators, which will prove to be useful tools if our system does not supply such a function. Finally, we discussed the importance of evaluating the reasonableness of the generated random numbers and we saw two programs that will carry out that evaluation.

In the next chapter we will continue to expand our understanding of BASIC; we will see how to express vectors and matrices in the form of arrays.

EXERCISES

2-1: Alter the program given in Figure 2.3 so that the number of items to be averaged is determined by the program. Use a value less than − 20000 as a signal for the end of data. The program should repeatedly loop, receiving items and incrementing the number of items, N%, until a number less than − 20000 is entered.

2-2: Alter the program given in Figure 2.3 so that the items to be averaged are obtained from a **DATA** statement in the program rather than from the keyboard. Use the value − 20000 as a signal for the end of data. The program should repeatedly loop, receiving items and incrementing the number of items, N%, until the number − 20000 is encountered.

2-3: Alter the program given in Figure 2.3 so that all values entered from the console are read as strings. Inspect the values for proper range. Convert each value to the corresponding numeric quantity if it is in the correct range, or print an error message if it is out of range.

2-4: Verify that the first fifteen digits of pi have an average value of 5.1 and a standard deviation of 2.8.

2-5: Alter the program given in Figure 2.8 so that 24 uniformly distributed random numbers are used to generate each Gaussian random number.

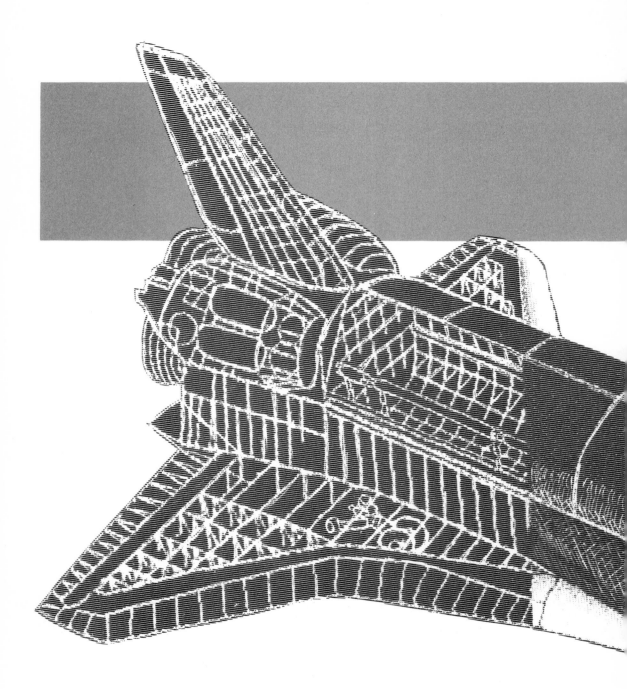

CHAPTER 3

Vector and Matrix Operations

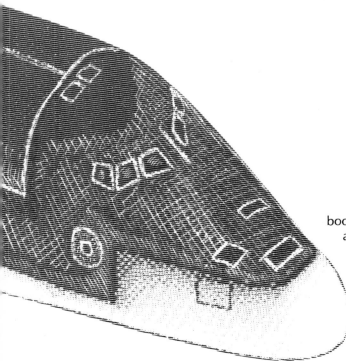

INTRODUCTION

In most of the chapters of this book, we develop programs that utilize vectors and matrices. Consequently, in this chapter, we will review the concepts of vectors and matrices and consider some of their more common mathematical operations. We will demonstrate several BASIC implementations. Vectors and matrices are important concepts because they greatly simplify the programming of mathematical operations on data sets.

SCALARS AND ARRAYS

We will begin our discussion by establishing the difference between a *scalar* variable and an *array*. An ordinary, simple variable is called a scalar. It is referenced by a unique symbolic name, and it is associated with a single value. For example, the BASIC expression:

YEAR = 1971

assigns a value to the scalar variable YEAR. Certain dialects of BASIC also include the integer scalar type. A percent-sign is used as a suffix in this case, for example:

YEAR% = 1971

Sometimes it is necessary to refer collectively to a set of scalar values. BASIC utilizes the *array* for this purpose. In an array, all of the components have the same type. That is, all of the elements are real numbers, integers, or strings of characters.

The components of an array are collectively referenced by a symbolic name. Each position of the array is uniquely designated by an index or subscript that follows the name. The corresponding value at each position in the array can be individually accessed through this index. The value can be changed without affecting the other values of the array.

In the following sections we will see how *vectors* and *matrices* are represented as arrays in BASIC programs. We will begin by describing vectors.

VECTORS

A vector is a one-dimensional array. (The number of dimensions refers to the number of subscripts, not the number of elements.) Each element of a vector is referenced through a single index. In BASIC, the index begins with zero. Thus, if a BASIC vector contains 10 elements, the index number runs from zero through 9. The initial element, zero, is not often used. Therefore, we commonly refer to the element referenced by the index of unity as the first element of the vector.

In ordinary usage, the elements of a vector are separated by spaces. For example, consider the vector **v** that contains the values:

2 5 1 9 4 3

The BASIC statement:

DIM A(5), V%(6)

defines an array A of real numbers and an array V of integers. The maximum number of elements (the length) is declared to be 6 for array A and 7 for array V. (The value in parentheses refers to the largest index number, which is one less than the number of elements.) Values can be assigned to arrays by BASIC statements such as:

$$V\%(1) = 2$$
$$V\%(2) = 5$$
$$V\%(3) = 1$$
$$V\%(4) = 9$$
$$V\%(5) = 4$$
$$V\%(6) = 3$$

In this example, the zeroth element, $V\%(0)$, is not defined. The first element of this vector is located at position $V\%(1)$ and has a value of 2, the second element is at $V\%(2)$ and has a value of 5, and the last element is at $V\%(6)$ and has a value of 3.

Since vectors have only a single dimension, it should not matter whether they are written horizontally or vertically. But sometimes we will need to distinguish between *row vectors*, which are written horizontally, and *column vectors*, which are written vertically. In this case, the set:

$$\begin{bmatrix} 2 \\ 5 \\ 1 \\ 9 \\ 4 \\ 3 \end{bmatrix}$$

is a column vector.

In the next section we will study the operations of vector arithmetic and their implementation in BASIC.

Vector Arithmetic

The major arithmetic operations defined for vectors are: magnitude, scalar multiplication, vector addition, dot product, and cross product. Let us now look at each of these operations.

Magnitude

The *magnitude* of a vector is a scalar value. It is obtained by summing the squares of the elements, then taking the square root of the resulting sum. For example, the magnitude of the vector **y**:

$$\mathbf{y} = \begin{bmatrix} 2 & 2 & 1 \end{bmatrix}$$

is equal to the square root of $4 + 4 + 1$. The resulting value is 3. The operation can be programmed with the BASIC statement:

```
M3 = SQR(Y(1) * Y(1) + Y(2) * Y(2) + Y(3) * Y(3))
```

Scalar Multiplication of Vectors

If a vector **y** is multiplied by a scalar value *s*, then each element of **y** is multiplied by *s*. For example, 2 times the vector **y** is:

$$2\mathbf{y} = \begin{bmatrix} 4 & 4 & 2 \end{bmatrix}$$

The following BASIC expression generates an array Y2 in which each element is twice as large as the corresponding element of array Y:

```
FOR I = 1 TO 3
    Y2(I) = 2.0 * Y(I)
NEXT I
```

Vector Addition

Two vectors can be added together if each has the same number of elements, or the same length. The result is a new vector in which each element is the sum of the two corresponding elements of the original vectors. Thus if:

$$\mathbf{a} = \begin{bmatrix} 1 & 2 & 3 \end{bmatrix}$$

and

$$\mathbf{b} = \begin{bmatrix} 3 & 4 & 5 \end{bmatrix}$$

then

$$\mathbf{a} + \mathbf{b} = \begin{bmatrix} 4 & 6 & 8 \end{bmatrix}$$

The corresponding BASIC expression is:

FOR I = 1 **TO** 3
    C(I) = A(I) + B(I)
NEXT I

Dot Product (or Scalar Product)

The *dot product* or *scalar product* of two equal-length vectors produces a scalar result. Each element of one vector is multiplied by the corresponding element of the other. The resulting products are then added together. The mathematical symbol for the dot operator is simply a dot placed between the operands. Thus:

$$\mathbf{a} \cdot \mathbf{b} = (1)(3) + (2)(4) + (3)(5) = 26$$

The dot product is equal to the product of the magnitude of the original vectors and the cosine of the angle between them:

$$\mathbf{a} \cdot \mathbf{b} = |\mathbf{a}|\,|\mathbf{b}|\cos\theta$$

This formula can be used to find the angle between two vectors. For example, the angle between the vectors **a** and **b** is 10.7°:

$$\cos\theta = 26/(3.742 \cdot 7.071) = 0.9826 \quad \text{and}$$
$$\text{Arccos } 0.9826 = 10.7°$$

Cross Product (or Vector Product)

The *cross product*, or *vector product*, of two vectors produces a third vector that is mutually perpendicular to the two original vectors. The mathematical symbol for this operation is an ×. The magnitude of the resulting vector is equal to the product of the magnitudes of the original vectors and the sine of the angle between them:

$$|\mathbf{a} \times \mathbf{b}| = |\mathbf{a}|\,|\mathbf{b}|\sin\theta$$

The cross product, $\mathbf{c} = \mathbf{a} \times \mathbf{b}$, can be calculated from the BASIC expressions:

C(1) = A(2) * B(3) − B(2) * A(3)
C(2) = −A(1) * B(3) + B(1) * A(3)
C(3) = A(1) * B(2) − B(1) * A(2)

The cross product of the vectors **a** and **b** is the vector $\begin{bmatrix} -2 & 4 & -2 \end{bmatrix}$; its magnitude is 4.9.

We have now seen BASIC implementations for the main arithmetic operations on vectors. Later in this chapter we will see the matrix equivalents of these operations. However, before moving on to matrices we should define one special kind of vector implementation—the *string*.

Strings

Strings of alphabetic and numeric (alphanumeric) characters are useful for representing such things as names and addresses. These can easily be handled in BASIC by defining a vector of strings. That is, each name or address is one element of the vector. In BASIC, a string is indicated by a dollar-sign suffix. The declaration:

DIM NAME$(100)

defines a vector of 101 strings, since the index can range from zero to 100. Each of the 101 elements can contain a string of alphanumeric characters.

Strings are defined by enclosing the characters in quotation marks. For example, the statement:

NAME$(1) = "John Smith"

initializes the first string element. Sorting and other operations can be performed on strings:

IF NAME$(I) < NAME$(J) **THEN** . . .

Now that we have described vectors and strings, let us move on to two-dimensional arrays and their BASIC representations.

MATRICES

A two-dimensional array is called a *matrix*. The elements of this set are arranged into a rectangle or square. The elements of the matrix can be considered as a set of horizontal lines called row vectors or as a set of vertical lines called column vectors. Thus, a matrix can be described as a one-dimensional set of vectors.

Each element of a matrix is uniquely defined by a pair of indices: the *row index* and the *column index*. For example, consider the matrix:

$$\begin{bmatrix} x_{11} & x_{12} & x_{13} & \cdots & x_{1m} \\ x_{21} & x_{22} & x_{23} & \cdots & x_{2m} \\ x_{31} & x_{32} & x_{33} & \cdots & x_{3m} \\ \cdots & \cdots & \cdots & \cdots & \cdots \\ x_{n1} & x_{n2} & x_{n3} & \cdots & x_{nm} \end{bmatrix}$$

which contains n rows and m columns. The row index is always given first. It is then followed by the column index. Note that the zeroth row and column are not shown.

A matrix is referenced by its name, which can be a single alphabetic character, or a string of characters. The indices are given as subscripts except in computer programs, where subscripting is not possible. In BASIC, FORTRAN, and COBOL programs, the array subscripts are enclosed in parentheses. Square brackets are used for this purpose in Pascal and APL. Thus, the appearance of the expression X(2,3) in a BASIC program is a reference to the element located in row two and column three of the two-dimensional array named X.

A matrix that has the same number of rows as columns is called a *square matrix*. The *principal*, or *main diagonal* of a square matrix contains the elements x_{11}, x_{22}, x_{33}, . . . , x_{nn}. The principal diagonal is sometimes referred to simply as the *diagonal*. A square matrix that contains the value of unity at each position of the main diagonal, and is zero everywhere else, is known as a *unit* matrix, or an *identity* matrix. For example:

$$\begin{bmatrix} 1 & 0 & 0 & 0 \\ 0 & 1 & 0 & 0 \\ 0 & 0 & 1 & 0 \\ 0 & 0 & 0 & 1 \end{bmatrix}$$

is a 4-by-4 unit matrix.

Now that we have defined the essential vocabulary of matrices, we can go on to study the major arithmetic operations for matrices, and the BASIC implementations of these operations.

Matrix Arithmetic

We will begin by defining the transpose operation; then we will describe scalar multiplication, and matrix addition, subtraction, and multiplication.

The Transpose Operation

A matrix is *transposed* by interchanging the rows and the columns. Each original element x_{ij} becomes the new element x_{ji} of the transposed matrix. The *transpose* of matrix X is designated as X^T. Thus, if

$$X = \begin{bmatrix} 1 & 2 & 3 \\ 4 & 5 & 6 \\ 7 & 8 & 9 \end{bmatrix}$$

then

$$X^T = \begin{bmatrix} 1 & 4 & 7 \\ 2 & 5 & 8 \\ 3 & 6 & 9 \end{bmatrix}$$

Notice that the transpose of a square matrix can be obtained by rotation of the matrix about the principal diagonal.

Two matrices are *equal* if every element of one is equal to the corresponding element of the other. Thus if $X = Y$, then for each element:

$$x_{ij} = y_{ij}$$

A square matrix is symmetric if it is equal to its transpose. In this case, each element x_{ij} equals the corresponding element x_{ji}.

Scalar Multiplication of Matrices

If a matrix X is multiplied by a scalar value s, then each element of the matrix is multiplied by the value s. The following BASIC statements will produce an array Y from the product of array X and the scalar S.

```
FOR I% = 1 TO N%
   FOR J% = 1 TO M%
      Y(I%,J%) = X(I%,J%) * S
   NEXT J%
NEXT I%
```

Matrix Addition and Subtraction

One matrix may be added to or subtracted from another matrix if both have the same number of columns and the same number of rows. The addition of the matrix X to the matrix Y to produce matrix Z is written as:

$$Z = X + Y$$

Thus, if

$$X = \begin{bmatrix} x_{11} & x_{12} & x_{13} \\ x_{21} & x_{22} & x_{23} \end{bmatrix}$$

and

$$Y = \begin{bmatrix} y_{11} & y_{12} & y_{13} \\ y_{21} & y_{22} & y_{23} \end{bmatrix}$$

then

$$Z = \begin{bmatrix} x_{11} + y_{11} & x_{12} + y_{12} & x_{13} + y_{13} \\ x_{21} + y_{21} & x_{22} + y_{22} & x_{23} + y_{23} \end{bmatrix}$$

The corresponding BASIC statements are:

```
FOR I% = 1 TO N%
  FOR J% = 1 TO M%
    Z(I%,J%) = X(I%,J%) + Y(I%,J%)
  NEXT J%
NEXT I%
```

Each element of the array Z is formed from the sum from the corresponding elements of X and Y. In a similar way, subtraction of one matrix from another:

$$Z = X - Y$$

is performed by subtracting each element of the second matrix from the corresponding element of the first.

Matrix Multiplication

One matrix may be multiplied by another if the number of columns of the first matrix equals the number of rows of the second. The two matrices are said to be *conformable* in this case. Thus, if X is a matrix that contains m rows and n columns, and Y is a matrix with n rows and p columns, then the product:

$$Z = X Y$$

produces a matrix Z with m rows and p columns. That is, the resulting matrix has the same number of rows as matrix X and the same number of columns as matrix Y.

In matrix multiplication, each element of Z is formed from a sum of products. The elements from one row of matrix X are each multiplied by the corresponding elements from a column of matrix Y, then summed up. For matrix X (which might be the transpose of the previous X) and matrix Y:

$$X = \begin{bmatrix} x_{11} & x_{12} \\ x_{21} & x_{22} \\ x_{31} & x_{32} \end{bmatrix}$$

and

$$Y = \begin{bmatrix} y_{11} & y_{12} & y_{13} \\ y_{21} & y_{22} & y_{23} \end{bmatrix}$$

the operation $Z = XY$ is formed as

$$Z = \begin{bmatrix} x_{11}y_{11} + x_{12}y_{21} & x_{11}y_{12} + x_{12}y_{22} & x_{11}y_{13} + x_{12}y_{23} \\ x_{21}y_{11} + x_{22}y_{21} & x_{21}y_{12} + x_{22}y_{22} & x_{21}y_{13} + x_{22}y_{23} \\ x_{31}y_{11} + x_{32}y_{21} & x_{31}y_{12} + x_{32}y_{22} & x_{31}y_{13} + x_{32}y_{23} \end{bmatrix}$$

Each jk element of Z is formed from row j of matrix X and column k of matrix Y according to the scheme

$$z_{jk} = x_{j1}y_{1k} + x_{j2}y_{2k} + \ldots + x_{jn}y_{nk}$$

Thus, if

$$X = \begin{bmatrix} 1 & 4 \\ 2 & 5 \\ 3 & 6 \end{bmatrix}$$

and

$$Y = \begin{bmatrix} 7 & 8 & 9 \\ 10 & 11 & 12 \end{bmatrix}$$

then the product:

$$Z = XY = \begin{bmatrix} 47 & 52 & 57 \\ 64 & 71 & 78 \\ 81 & 90 & 99 \end{bmatrix}$$

is a square matrix with dimensions of 3 by 3. The first element, z_{11}, for example, is calculated as:

$$(1)(7) + (4)(10) = 47$$

Matrix multiplication is not commutative; that is, the product YX will not, in general, be equal to the product XY. Reversing the order of the previous example produces a 2-by-2 matrix, rather than a 3-by-3 matrix:

$$YX = \begin{bmatrix} 50 & 122 \\ 68 & 167 \end{bmatrix}$$

Notice that the dot product of two vectors follows the rules for matrix multiplication if the first vector is a column vector (that is, has one column), and the second is a row vector (that is, has one row).

We have seen how BASIC handles vector and matrix arithmetic through the use of one- and two-dimensional arrays. We are now ready to write a program using a number of the BASIC statements we have learned.

BASIC PROGRAM: MATRIX MULTIPLICATION

In Chapter 4, we will need a routine for matrix multiplication. Specifically, we will need to multiply the transpose of a matrix U by the original matrix U. We will also need to multiply the vector **y** by the matrix U. Therefore, we will program such a routine now.

The program shown in Figure 3.1 contains a subroutine that will perform both of the needed operations. The subroutine at line 500 will generate the U matrix and the **y** vector. The subroutine at line 4000 will calculate the needed matrix A and vector **z** according to the equations:

$$U^T U = A$$

and

$$yU = z$$

The matrix U, in this case, contains 5 rows and 3 columns. Vector **y** has a length of 5.

The main program assigns values to N1% and N2%, dimensions the four arrays, and calls the two subroutines. It also creates the strings A\$ and B\$ (lines 30 and 40), which will be used as a formatting template for the **PRINT USING** statements. After the subroutine calls, the main program prints out both the original matrix U and vector **y** and the calculated matrix A and vector **z**.

The nested **FOR** loop is an ideal tool for handling matrices. Consider the loops:

FOR I% = 1 **TO** N1%

  . . .

  FOR J% = 1 **TO** N2%

    . . .

  NEXT J%

  . . .

NEXT I%

For each incrementation of I% in the outer loop, the lines of the inner loop are executed N2% times. This is a very economical way to print a matrix, as we can see in the main program. Notice the punctuation of lines 160 and 180. The semicolon at the end of line 160 suppresses the automatic line feed and causes each element of a row of matrix U to be printed on the same line. On the other hand, the absence of punctuation at the end of line 180 generates a new line after the element of vector **y** has been printed.

Nested loops occur again in the subroutine at line 500. Line 550 assigns the value of unity to each element of the first column of matrix U. Line 570 calculates the other two columns: column 2 receives the values 1 through 5; column 3 is calculated as $1 \cdot 1$ to $5 \cdot 5$.

The subroutine at line 4000 merits careful study. It contains three levels of **FOR** loops, and illustrates the elegance of nesting. The innermost loop (lines 4050 to 4080) performs the algorithm for matrix multiplication (or, more precisely, the multiplication of a matrix by its transpose). There are two second-level loops (lines 4030 to 4090 and 4110 to 4130). Line 4030 takes into account that the matrix A will be symmetric; only the elements of the diagonal and above need be calculated. The elements below the diagonal are assigned by line 4070. There will be terms like:

$$A(1,3) = A(3,1)$$

The loop at lines 4110 to 4130 performs the multiplication yU.

Notice line 4040:

4040 A(K%,L%) = 0

This line initializes each element of A to zero before the summation process begins in line 4060. This step may not be necessary since some BASICs initialize all variables to zero when program execution begins. However, if we were to revise this program to call the subroutine at line 4000 more than once, line 4040 would be essential. Without it, the summation of each element of A would begin at whatever value was left over from the previous subroutine call. Thus, the initialization in line 4040 is good programming practice.

Matrix A will have 3 rows (the same as the transpose of matrix U) and 3 columns (the same as matrix U). Vector **z** will have a length of 3. Actually, both **y** and **z** must be considered as row vectors; that is, they have a dimension of 1 row and 5 columns. Alternately, if we want to consider the vectors **y** and **z** as column vectors, then we should write the multiplication equation as:

$$\mathbf{y}^T U = \mathbf{z}^T$$

where the transposed column vectors become row vectors.

Type up the program shown in Figure 3.1 and execute it. The results should look like Figure 3.2.

```
10    REM Matrix multiplication, Apr 19, 81
11    REM identifiers
13    REM      N1%    NROW%      number of rows
14    REM      N2%    NCOL%      number of columns
15    REM end of identifiers
20    REM
30    A$ = '' ####.# ''
40    B$ = ''  = ####.#''
50    N1% = 5
60    N2% = 3
70    DIM Z(3), A(3,3), Y(5), U(5,3)
80    REM
90    GOSUB 500 : REM get the data
100   GOSUB 4000 : REM square up the matrix
110   REM
120   PRINT
130   PRINT ''                  U                    Y''
```

Figure 3.1: Matrix Multiplication, $U^T U = A$, $Y U = Z$.

```
140    FOR I% = 1 TO N1%
150        FOR J% = 1 TO N2% : REM original data
160            PRINT USING A$; U(I%,J%);
170        NEXT J%
180        PRINT USING B$; Y(I%)
190    NEXT I%
200    PRINT
210    PRINT "                A                    Z"
220    FOR I% = 1 TO N2%
230        FOR J% = 1 TO N2% : REM original data
240            PRINT USING A$; A(I%,J%);
250        NEXT J%
260        PRINT USING B$; Z(I%)
270    NEXT I%
280    PRINT
290    GOTO 9999
500    REM
510    REM input the data
520    REM
530    PRINT
540    FOR I% = 1 TO N1%
550        U(I%,1) = 1
560        FOR J% = 2 TO N2%
570            U(I%,J%) = I% * U(I%,J%-1)
580        NEXT J%
590        Y(I%) = 2 * I%
600    NEXT I%
4000   REM U and Y converted to A and Z
4011   REM identifiers
4013   REM      N1%    NROW%      number of rows
4014   REM      N2%    NCOL%      number of columns
4015   REM end of identifiers
4016   REM
4020   FOR K% = 1 TO N2%
4030       FOR L% = 1 TO K%
```

Figure 3.1: Matrix multiplication, $U^T U = A$, $Y U = Z$ (cont.)

```
4040        A(K%,L%) = 0
4050        FOR I% = 1 TO N1%
4060            A(K%,L%) = A(K%,L%) + U(I%,L%)*U(I%,K%)
4070            IF (K% <> L%) THEN A(L%,K%) = A(K%,L%)
4080        NEXT I%
4090        NEXT L%
4100        Z(K%) = 0
4110        FOR I% = 1 TO N1%
4120            Z(K%) = Z(K%) + Y(I%) * U(I%,K%)
4130        NEXT I%
4140    NEXT K%
4150    RETURN : REM from square
9999    END
```

Figure 3.1: Matrix multiplication, $U^T U = A$, $Y U = Z$ (cont.)

```
               U                         Y
      1.0     1.0        1.0   =       2.0
      1.0     2.0        4.0   =       4.0
      1.0     3.0        9.0   =       6.0
      1.0     4.0       16.0   =       8.0
      1.0     5.0       25.0   =      10.0

               A                         Z
      5.0    15.0       55.0   =      30.0
     15.0    55.0      225.0   =     110.0
     55.0   225.0      979.0   =     450.0
```

Figure 3.2: Output from the Matrix Multiplication Program

DETERMINANTS

The *determinant* of a square matrix X is designated as $|X|$. The result is a scalar value. For a 2-by-2 matrix, the upper-left member is multiplied by the lower-right, and then the product of the lower-left member and the upper-right member is subtracted:

$$|X| = x_{11}x_{22} - x_{12}x_{21}$$

For example, the determinant of the matrix:

$$\begin{bmatrix} 1 & 2 \\ 3 & 4 \end{bmatrix}$$

is -2.

The determinant of matrices larger than 2 by 2 can be found by multiplying each element of the first row by the determinant of the remaining matrix, after removing the column that is common to the element in the first row. A recursive definition is:

$$|X| = x_{11}s_{11} - x_{12}s_{12} + x_{13}s_{13} - \ldots (-1)^{n+1}x_{1n}s_{1n}$$

where x_{11}, x_{12}, etc., are the elements of the first row of matrix X and s_{1n} is the determinant of the matrix that has row 1 and column n removed. The determinant of the matrix:

$$\begin{bmatrix} 1 & 2 & 3 \\ 4 & 5 & 6 \\ 7 & 8 & 0 \end{bmatrix}$$

is equal to:

$$1 \begin{vmatrix} 5 & 6 \\ 8 & 0 \end{vmatrix} - 2 \begin{vmatrix} 4 & 6 \\ 7 & 0 \end{vmatrix} + 3 \begin{vmatrix} 4 & 5 \\ 7 & 8 \end{vmatrix}$$

The next step is to evaluate the minor 2-by-2 matrices:

$$\begin{vmatrix} 5 & 6 \\ 8 & 0 \end{vmatrix} \quad \begin{vmatrix} 4 & 6 \\ 7 & 0 \end{vmatrix} \quad \begin{vmatrix} 4 & 5 \\ 7 & 8 \end{vmatrix}$$

according to the procedure:

$$1(5 \cdot 0 - 8 \cdot 6) - 2(4 \cdot 0 - 7 \cdot 6) + 3(4 \cdot 8 - 7 \cdot 5)$$

The resulting value of the determinant, in this case, is 27. If each element of a row or column is zero, then the determinant is zero. Also, if two rows or columns are identical, then the determinant is zero.

We have described the method for calculating the determinant of a matrix. Since we will be using determinants in later chapters, let us now consider a BASIC program that calculates determinants.

BASIC PROGRAM: DETERMINANTS

A program that can be used to find the determinant of a 3-by-3 matrix is given in Figure 3.3. The input subroutine, located at line 500, asks the user to enter the matrix elements. The subroutine at line 5280 then calculates the determinant. Type up the program and run it. The elements of the matrix are entered row by row. That is, the order is:

$$A(1,1), A(1,2), A(1,3), A(2,1), A(2,2), \ldots$$

Enter the matrix from the previous section and verify that the result is 27. After the nine elements have been entered, the program displays the value of the determinant. The cycle then repeats.

```
 10   REM Calculation of a 3—by—3 determinant, Apr 14, 81
 11   REM identifiers
 16   REM      N1%    NROW%      number of rows
 17   REM      N2%    NCOL%      number of columns
 18   REM end of identifiers
 20   REM
 30   A$ = " ###.### "
 40   N1% = 3
 50   N2% = N1%
 60   DIM B(3,3)
 70   REM
 80   PRINT " The determinant of a 3—by—3 matrix"
 90   GOSUB 500 : REM input subroutine
100   GOSUB 5280 : REM Cramer's rule
110   PRINT
120   FOR I% = 1 TO N1%
130     FOR J% = 1 TO N2%
140       PRINT USING A$; B(I%,J%);
150     NEXT J%
160   PRINT
170   NEXT I%
180   PRINT
190   PRINT " Determinant = "; SUM
200   PRINT
210   GOTO 90 : REM next set
500   REM
510   REM Input subroutine
520   REM
530   PRINT
540   FOR I% = 1 TO N1%
550     PRINT "Row"; I%
560     FOR J% = 1 TO N1%
570       PRINT J%; "  ";
580       INPUT B(I%,J%)
590     NEXT J%
```

Figure 3.3: The Determinant of a 3-by-3 Matrix

```
600   NEXT I%
610   RETURN : REM from input subroutine
5280  REM Find the determinant of a 3—by—3 matrix
5290  REM
5300  SUM = B(1,1) * (B(2,2) * B(3,3) — B(3,2) * B(2,3))
5310  SUM = SUM — B(1,2) * (B(2,1) * B(3,3) — B(3,1) * B(2,3))
5320  SUM = SUM + B(1,3) * (B(2,1) * B(3,2) — B(3,1) * B(2,2))
5330  RETURN : REM from determinant subroutine
9999  END
```

Figure 3.3: The determinant of a 3-by-3 matrix (cont.)

INVERSE MATRICES AND MATRIX DIVISION

The *inverse* of a nonsingular matrix X is written as

$$X^{-1}$$

The inverse of a singular matrix is undefined. The product of a matrix and its inverse is the identity matrix:

$$XX^{-1} = I$$

Matrix inversion is similar to other inverse operations since the inverse of an inverse produces the original matrix:

$$(X^{-1})^{-1} = X$$

In computer implementations, however, the two will not always agree because of roundoff error.

Matrix division is performed by a combination of inversion and multiplication. Thus, if we need to divide matrix X by matrix Y, we first perform a matrix inversion on Y. Then matrix X is multiplied by the inverse of matrix Y. The operation is written as:

$$XY^{-1}$$

Matrix inversion can be used to find the solution to a set of simultaneous linear equations. Thus, if we have a coefficient matrix A and a constant vector **y**, we want a solution vector **b** such that

$$A\mathbf{b} = \mathbf{y}$$

Then the solution can be obtained from a product of the inverse of matrix A and the constant vector **y** (in that order).

$$A^{-1}\mathbf{y} = \mathbf{b}$$

Since the simultaneous solution of linear equations is the subject of the next chapter, we shall not develop a matrix inversion routine at this point.

SUMMARY

In this chapter we saw how vectors and matrices are represented in BASIC. We also studied the methods of performing matrix and vector arithmetic operations in BASIC. We developed two significant programs—one for performing matrix multiplication, and the other for calculating determinants. We will be using these programs in the chapters that follow.

EXERCISES

3-1: *Write a program that will input two three-element vectors from the keyboard and print the corresponding dot product. Show that the dot product of the vectors:*

$a = [1\ 1\ 1]$ *and* $b = [1\ -1\ -1]$

is -1.

3-2: *Write a program that will input two three-element vectors from the keyboard and print the corresponding cross product. Show that the cross product of the vectors:*

$a = [1\ 1\ 1]$ *and* $b = [1\ -1\ 0]$

is the vector $[1\ 1\ -2]$.

3-3: *Write a program that will input two three-element vectors from the keyboard and print the corresponding angle between the two vectors. Show that the angle between the vectors:*

$a = [1\ 1\ 1]$ *and* $b = [1\ -1\ -1]$

is 109.5 degrees. (This is the O-Si-O bond angle in silicate minerals.)

3-4: *A unit vector has a magnitude of unity. Dividing a vector by its magnitude produces a unit vector. Write a program to convert a three-dimensional vector, input from the keyboard, into a unit vector. Run the program to show that the vector:*

$a = [18\ -18\ 9]$

is associated with the unit vector $[2/3\ -2/3\ 1/3]$.

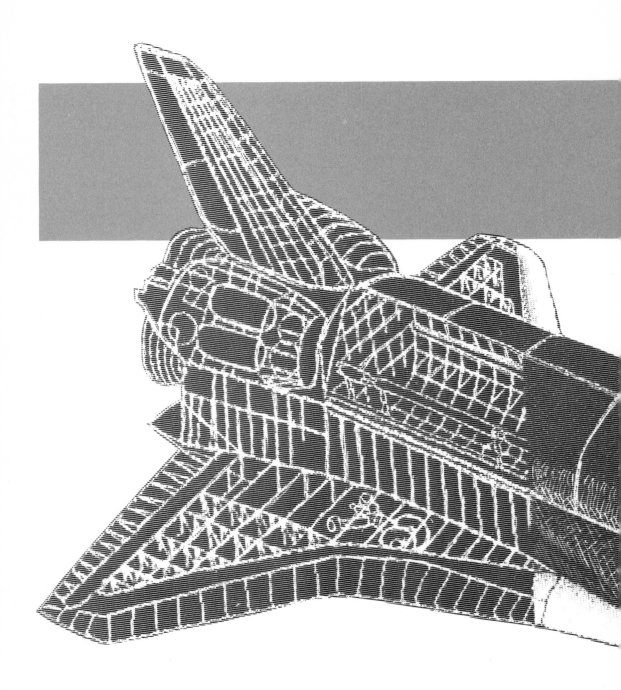

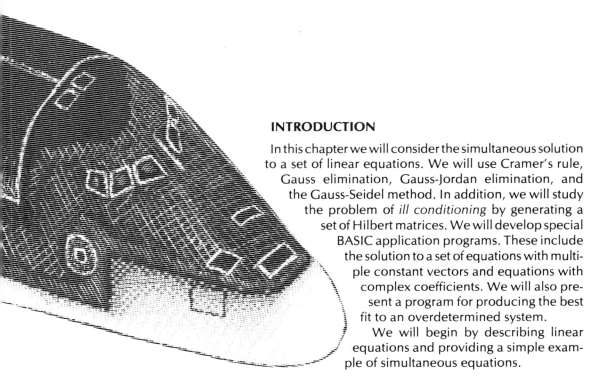

CHAPTER **4**

Simultaneous Solution of Linear Equations

INTRODUCTION

In this chapter we will consider the simultaneous solution to a set of linear equations. We will use Cramer's rule, Gauss elimination, Gauss-Jordan elimination, and the Gauss-Seidel method. In addition, we will study the problem of *ill conditioning* by generating a set of Hilbert matrices. We will develop special BASIC application programs. These include the solution to a set of equations with multiple constant vectors and equations with complex coefficients. We will also present a program for producing the best fit to an overdetermined system.

We will begin by describing linear equations and providing a simple example of simultaneous equations.

LINEAR EQUATIONS AND SIMULTANEOUS SOLUTIONS

A *linear equation* consists of a sum of terms such as:

$$Ax + By + Cz = D$$

In this equation, x, y and z are variables and A, B, C, and D are constants. No more than one variable can occur in each term, and it must be present to the first power. Thus, expressions such as:

$$2x^2 + 3y^2 = 4$$
$$\sin(x) + \log(y) = 9$$

and

$$xy = 2$$

are *nonlinear equations* if the variables are x and y.

An equation such as:

$$\frac{x}{A} + y \log(B) = p$$

is linear in the parameters x and y. On the other hand, if the symbols A and B are considered as the variables, and the symbols x, y and p are known values, then the equation is nonlinear.

Linear equations occur frequently in all branches of science and engineering and so effective methods for their solution are necessary. For the particular case where there are several unknowns and an equal number of independent equations, a unique solution is possible. In this chapter, we will consider several different methods for the solution of a set of linear equations. We will discuss the solution of nonlinear equations in Chapter 10.

The two linear equations:

$$x - 2y = 1$$
$$2x + y = 7$$

represent two straight lines in the x-y plane. The simultaneous solution of these two equations is the intersection of the two lines. Therefore, a graphical solution can be obtained by plotting the lines and finding the point of intersection.

The first equation has a y-intercept of -0.5 and a slope of 0.5. This can be seen by rearranging the equation as:

$$y = -0.5 + 0.5x$$

The second equation has an intercept of 7 and a slope of -2:

$$y = 7 - 2x$$

The intersection of the two lines occurs at the location:

$$x = 3, \quad y = 1$$

which represents the solution to the problem. This solution can be verified by substitution into the original equations.

Another method for finding the simultaneous solution to these two equations is to multiply the second equation by 2 and add it to the first equation. The resulting equation contains only the variable x, so it can be solved directly.

$$
\begin{array}{rcl}
x - 2y &=& 1 \\
4x + 2y &=& 14 \\
\hline
5x &=& 15 \quad \text{or} \quad x = 3
\end{array}
$$

This value of x can then be substituted into either one of the original equations to find the corresponding value of y. Both of the above methods are suitable for solving two simultaneous equations. But, in general, these methods are tedious for larger numbers of equations. In the next section we will study a more sophisticated method of solving simultaneous equations, using matrices and vectors.

SOLUTION BY CRAMER'S RULE

A technique known as Cramer's rule is useful for solving two or three simultaneous equations. This is a particularly powerful method when the solution is performed by hand, or when a pocket calculator is used. In this approach, the equations are written with the unknowns on one side of the equal sign, and the constant terms on the other. Terms containing the same unknowns are vertically aligned. The two equations in the previous section were initially written in this form.

The coefficients of the unknowns are placed into a matrix known as the *coefficient matrix*. The constant terms are put into a separate vector. For our two equations, this produces the matrix A and the vector $\mathbf{z}$.

$$
A = \begin{bmatrix} 1 & -2 \\ 2 & 1 \end{bmatrix} \qquad \mathbf{z} = \begin{bmatrix} 1 \\ 7 \end{bmatrix}
$$

The solution is found from the relationship:

$$x = \frac{D_1}{D} \qquad y = \frac{D_2}{D}$$

where D is the determinant of the coefficient matrix:

$$
D = \begin{vmatrix} 1 & -2 \\ 2 & 1 \end{vmatrix} = (1)(1) - (2)(-2) = 5
$$

The determinant D_1 is found by substituting the constant vector into column 1 of the matrix. This column corresponds to the unknown x:

$$D_1 = \begin{vmatrix} 1 & -2 \\ 7 & 1 \end{vmatrix} = (1)(1) - (7)(-2) = 15$$

Similarly, D_2 is obtained by substituting the constant vector into column 2:

$$D_2 = \begin{vmatrix} 1 & 1 \\ 2 & 7 \end{vmatrix} = (1)(7) - (2)(1) = 5$$

The solution is then:

$$x = \frac{15}{5} = 3 \quad \text{and} \quad y = \frac{5}{5} = 1$$

Of course, some sets of equations have no unique solutions. We will now examine how Cramer's rule handles such equations.

Linear Dependence

We can see that if the determinant of the coefficient matrix is zero, then no unique solution is possible. This will occur whenever there is a *linear dependence* among the equations; that is, one of the equations can be obtained from a combination of one or more of the others. The matrix is said to be singular in this case. As an example of linear dependence, consider the two equations:

$$\begin{aligned} x + y &= 5 \\ 2x + 2y &= 2 \end{aligned}$$

The coefficient matrix for these two equations is:

$$\begin{bmatrix} 1 & 1 \\ 2 & 2 \end{bmatrix}$$

and the corresponding determinant is zero. These two equations represent parallel lines, which, of course, do not intersect. Consequently, there can be no solution.

As a second example, consider the two equations:

$$\begin{aligned} x + y &= 5 \\ 2x + 2y &= 10 \end{aligned}$$

This example produces the same singular coefficient matrix as the previous one. In this case, however, the two lines lie on top of each other, and so there are an infinite number of solutions.

Now that we have a powerful method of solving two or three simultaneous equations, let us consider a practical application of the method. We will solve the problem posed in the following section using a BASIC program we developed in Chapter 3.

Example: A Direct-Current Electrical Circuit

An interesting practical example where simultaneous equations must be solved is provided by an electrical circuit. Consider, for example, the network of resistors and direct-current voltage sources shown in Figure 4.1. This circuit contains four nodes and six branches. There is a 20-volt source on the left side of the network, and a 5-volt source on the right side. In addition, there are six resistors of known value.

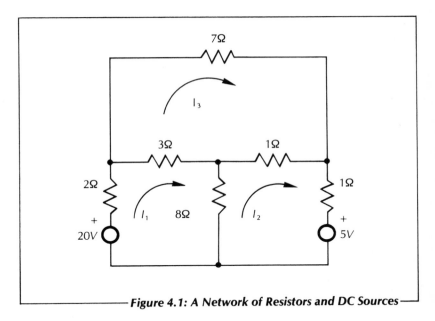

Figure 4.1: A Network of Resistors and DC Sources

The problem is to find the resulting branch currents and the voltages across each resistor. The electrical currents in the six separate branches can be determined by solving six simultaneous equations. The problem can be simplified, however, by considering three loop currents and the corresponding three simultaneous equations. Then, the branch currents can be found from the loop currents.

The loop current in the lower-left loop is designated as I_1, the loop current in the lower-right loop is I_2, and the loop current in the upper loop is I_3. Then the three loop equations can be derived from the Kirchhoff voltage law by going around each loop in a counter-clockwise direction:

$$13I_1 - 8I_2 - 3I_3 - 20 = 0 \text{ (lower-left loop)}$$
$$-8I_1 + 10I_2 - I_3 + 5 = 0 \text{ (lower-right loop)}$$
$$-3I_1 - I_2 + 11I_3 = 0 \text{ (upper loop)}$$

The corresponding coefficient matrix and constant vector are:

$$\begin{bmatrix} 13 & -8 & -3 \\ -8 & 10 & -1 \\ -3 & -1 & 11 \end{bmatrix} \qquad \begin{bmatrix} 20 \\ -5 \\ 0 \end{bmatrix}$$

Proceeding by Cramer's rule, we find:

$$D = \begin{vmatrix} 13 & -8 & -3 \\ -8 & 10 & -1 \\ -3 & -1 & 11 \end{vmatrix} \qquad D_1 = \begin{vmatrix} 20 & -8 & -3 \\ -5 & 10 & -1 \\ 0 & -1 & 11 \end{vmatrix}$$

$$D_2 = \begin{vmatrix} 13 & 20 & -3 \\ -8 & -5 & -1 \\ -3 & 0 & 11 \end{vmatrix} \qquad D_3 = \begin{vmatrix} 13 & -8 & 20 \\ -8 & 10 & -5 \\ -3 & -1 & 0 \end{vmatrix}$$

These determinants can be found by using the program given in Figure 3.3 of the previous chapter. The resulting loop currents are:

$$I_1 = \frac{D_1}{D} = \frac{1725}{575} = 3 \text{ amps}$$

$$I_2 = \frac{D_2}{D} = \frac{1150}{575} = 2 \text{ amps}$$

$$I_3 = \frac{D_3}{D} = \frac{575}{575} = 1 \text{ amp}$$

While this approach gives us the correct answer, it is unnecessarily complicated. Nine separate numbers must be entered into the computer program for each of the four determinants. This requires the entry of thirty-six separate values.

We will now study an efficient program designed specifically to make use of Cramer's Rule for solving simultaneous equations.

BASIC PROGRAM: A MORE ELEGANT USE OF CRAMER'S RULE

The program shown in Figure 4.3 simplifies the process considerably. The mathematical operations are exactly the same as in our first solution, but with this version, it is only necessary to enter twelve numbers: nine for the coefficients and three for the constant vector.

Running the Program

Type up the program shown in Figure 4.3 and execute it. Lines 5280-5330 of Figure 3.3 can be used directly. The coefficients and constant term for the first equation are entered in order. The RETURN key is pressed after each number is given. The corresponding values for the second equation are entered next and then the data for the third equation are entered. The input dialogue and the solution are shown in Figure 4.2.

```
    Simultaneous  solution  by  Cramer's  rule

    Equation  1
    1     ?  13
    2     ?  -8
    3     ?  -3
    C  ?  20
    Equation  2
    1     ?  -8
    2     ?  10
    3     ?  -1
    C  ?  -5
    Equation  3
    1     ?  -3
    2     ?  -1
    3     ?  11
    C  ?  0


    Matrix      Constants
    1.300E+01   -8.000E+00   -3.000E+00   =   2.000E+01
    -8.000E+00    1.000E+01   -1.000E+00   = -5.000E+00
    -3.000E+00   -1.000E+00    1.100E+01   =   0.000E+00
    Solution
    3.00000    2.00000    1.00000
```

Figure 4.2: Input/Output Dialogue: Solving Simultaneous Equations by Cramer's Rule

After the solution is displayed, the program will begin again. Another equation can be solved at this point.

```
10    REM Simultaneous solution of three linear equations
11    REM by Cramer's rule, Apr 14, 81
12    REM identifiers
14    REM      C1       COEF        Solution vector
15    REM      D3       DETERM      determinant
16    REM      E1%      ERMES%      error flag
17    REM      M1%      MAX%        maximum size
18    REM      N1%      NROW%       number of rows
19    REM      N2%      NCOL%       number of columns
20    REM      S6       SUM
21    REM end of identifiers
30    PRINT "Simultaneous solution  by Cramer's rule"
40    A$ = "  ##.###^^^ "
50    B$ = "  =  ##.###^^^ "
60    C$ = "   ##.#####"
70    M1% = 3
80    DIM Z(3), A(3,3), C1(3), B(3,3)
90    REM
100   GOSUB 500 : REM Input subroutine
110   GOSUB 5000 : REM Cramer's rule
120   REM
130   PRINT "        Matrix    Constants"
140   FOR I% = 1 TO N1%
150     FOR J% = 1 TO N2%
160       PRINT USING A$; A(I%,J%);
170     NEXT J%
180     PRINT USING B$; Z(I%)
190   NEXT I%
200   PRINT
210   IF (E1% = 1) THEN 280
220   PRINT "        Solution"
230   PRINT
```

Figure 4.3: Solution of Three Linear Equations by Cramer's Rule

```
240    FOR I% = 1 TO N2%
250        PRINT USING C$; C1(I%);
260    NEXT I%
270    PRINT
280    GOTO 100 : REM next set
500    REM
510    REM Input subroutine
520    REM
530    PRINT
540    N1% = 3
550    N2% = N1%
560    E1% = 0
570    FOR I% = 1 TO N1%
580        PRINT "Equation"; I%
590        FOR J% = 1 TO N1%
600            PRINT J%; "   ";
610            INPUT A(I%,J%)
620        NEXT J%
630        INPUT "  C "; Z(I%)
640    NEXT I%
650    RETURN : REM from input subroutine
5000   REM Solution of a 3—by—3 matrix by Cramer's rule
5010   REM Apr 14, 81
5011   REM identifiers
5013   REM      A        A           coefficient matrix
5014   REM      B        B           work array
5015   REM      C1       COEF        solution vector
5016   REM      D3       DETERM      determinant
5017   REM      E1%      ERMES%      error flag
5021   REM      N2%      NCOL%       number of columns
5022   REM      S6       SUM
5023   REM      Z        Z           constant vector
5024   REM end of identifiers
5070   REM
```

Figure 4.3: Solution of Three Linear Equations by Cramer's Rule (cont.)

```
5080    FOR I% = 1 TO N2%
5090       FOR J% = 1 TO N2%
5100          B(I%,J%) = A(I%,J%)
5110       NEXT J%
5120    NEXT I%
5130    GOSUB 5280 : REM find the determinant
5140    D3 = S6
5150    IF (D3 = 0) THEN 5250
5160    FOR J% = 1 TO N2%
5170       FOR I% = 1 TO N2%
5180          B(I%,J%) = Z(I%)
5190          IF (J% > 1) THEN B(I%,J%−1) = A(I%,J%−1)
5200       NEXT I%
5210       GOSUB 5280 : REM find the determinant
5220       C1(J%) = S6 / D3
5230    NEXT J%
5240    RETURN : REM normal return from Cramer's rule
5250    E1% = 1
5260    PRINT "ERROR—matrix singular "
5270    RETURN : REM from Cramer's rule
5280    REM Find the determinant of a 3—by—3 matrix
5290    REM
5300    S6 = B(1,1) * (B(2,2) * B(3,3) − B(3,2) * B(2,3))
5310    S6 = S6 − B(1,2) * (B(2,1) * B(3,3) − B(3,1) * B(2,3))
5320    S6 = S6 + B(1,3) * (B(2,1) * B(3,2) − B(3,1) * B(2,2))
5330    RETURN : REM from determinant subroutine
9999    END
```

Figure 4.3: Solution of Three Linear Equations by Cramer's Rule (cont.)

SOLUTION BY GAUSS ELIMINATION

Cramer's rule is an effective method for solving two or three simultaneous equations, particularly when the solution is performed by hand. However, the computation time increases with the fourth power of the matrix size. Consequently, it will take about sixteen times longer to solve six simultaneous equations than it does to solve three equations. The method of Gauss elimination is more efficient. Computation time increases with the third power of the number of equations, and so it will take about eight times longer to solve six equations than it does to solve three.

The Gauss elimination method can be readily programmed on a digital computer. However, the technique is fairly complicated, and therefore not suitable for hand solution. Furthermore, the operations involve frequent multiplication, division and subtraction. The consequent loss of precision places a practical upper limit on the number of equations that can be solved simultaneously. Let us go through the method step by step.

The Steps of the Gauss Method

With the Gauss elimination method, the original equations are manipulated so that the coefficient matrix contains a value of unity at each point on the major diagonal, and zero at each position below and to the left of the major diagonal.

Two basic types of matrix operations are used in the Gauss elimination method: scalar multiplication and addition. Any equation can be multiplied by a constant without changing the result. This is equivalent to multiplying one row of the coefficient matrix and the corresponding element in the constant vector by the same value. Also, any equation can be replaced by the sum of two equations.

The following steps will demonstrate the use of Gauss elimination using the equations we derived from the electric circuit shown in Figure 4.1. Initially, the coefficient matrix and the constant vector are:

$$\begin{bmatrix} 13 & -8 & -3 \\ -8 & 10 & -1 \\ -3 & -1 & 11 \end{bmatrix} \qquad \begin{bmatrix} 20 \\ -5 \\ 0 \end{bmatrix}$$

The first variable is eliminated from all but the first equation. The equations are manipulated to produce the value of unity at the top of the first column. The remaining positions of the column become zeros. The operation is performed in the following way. The entire first row is divided by the first element in the row (the pivot element). This generates the value of unity in the first diagonal position. By this means, the first equation becomes:

$$\begin{bmatrix} 1 & -0.61 & -0.23 \end{bmatrix} \qquad \begin{bmatrix} 1.5 \end{bmatrix}$$

The first unknown is eliminated from the second row by combining the first two rows. The new first row is multiplied by the first element in the second row, then subtracted from the second row. The new second row is:

$$\begin{bmatrix} 0 & 5.1 & -2.8 \end{bmatrix} \qquad \begin{bmatrix} 7.3 \end{bmatrix}$$

In a similar fashion, the first variable is eliminated from the third equation. The three equations are now approximately:

$$\begin{bmatrix} 1 & -0.61 & -0.23 \\ 0 & 5.1 & -2.8 \\ 0 & -2.8 & 10.3 \end{bmatrix} \qquad \begin{bmatrix} 1.5 \\ 7.3 \\ 4.6 \end{bmatrix}$$

The next step is to produce the value of unity in the second position of the second row (the new pivot). The second line is divided by the second element. The second variable is eliminated from the third equation by generating a zero in the second position, just under the pivot element. The three equations now look like:

$$\begin{bmatrix} 1 & -0.61 & -0.23 \\ 0 & 1 & -0.56 \\ 0 & 0 & 8.7 \end{bmatrix} \qquad \begin{bmatrix} 1.5 \\ 1.4 \\ 8.7 \end{bmatrix}$$

The final step of this phase is to obtain a value of unity in the third pivot position. This is accomplished by dividing the third equation by the pivot value to give:

$$\begin{bmatrix} 1 & -0.61 & -0.23 \\ 0 & 1 & -0.56 \\ 0 & 0 & 1 \end{bmatrix} \qquad \begin{bmatrix} 1.5 \\ 1.4 \\ 1 \end{bmatrix}$$

The result corresponds to the three equations:

$$x - 0.61y - 0.23z = 1.5$$
$$y - 0.56z = 1.4$$
$$z = 1$$

The third equation can be solved directly since it has only one unknown. The result is:

$$z = 1$$

The second equation becomes:

$$y = 1.4 + 0.56z$$

By substituting the value of z into this equation, we find the value of y to be 2. Substituting the values of y and z into the first equation produces the value of 3 for x. This phase of the calculations is known as *back substitution*.

Improving the Accuracy of the Gauss Method

The accuracy of the Gauss elimination method can be improved by interchanging two rows so that the element with the largest absolute magnitude becomes the pivot element. Suppose, for example, that the previous three equations had originally been written in a different order:

$$\begin{bmatrix} -3 & -1 & 11 \\ 13 & -8 & -3 \\ -8 & 10 & -1 \end{bmatrix} \qquad \begin{bmatrix} 0 \\ 20 \\ -5 \end{bmatrix}$$

Then the top equation could be divided by -3 to put the value of unity in the pivot position. However, the result will be more accurate if the first and second rows are interchanged first. This will put the larger value of 13 in the pivot position. After the first variable is eliminated from the second and third equations, the result is:

$$\begin{bmatrix} 1 & -0.61 & -0.23 \\ 0 & -2.8 & 10.3 \\ 0 & 5.1 & -2.8 \end{bmatrix} \qquad \begin{bmatrix} 1.5 \\ 4.6 \\ 7.3 \end{bmatrix}$$

Again, it would be best to interchange the second and third equations to put the larger element of 5.1 in the second pivot position.

There is another reason why rows may have to be interchanged. If a zero element appears on the major diagonal, it will not be possible to divide the row by this pivot element, but interchanging this row with one that is below it will remove the zero from the pivot position.

Now that we have gone through the rather tedious process of using Gauss elimination to solve equations by hand, we can appreciate the elegance of the BASIC program in the following section.

BASIC PROGRAM: THE GAUSS ELIMINATION METHOD

The program shown in Figure 4.4 can be used to solve a set of simultaneous linear equations by Gauss elimination. The program is written to handle up to eight equations. The size can be increased by changing the variable called M1% near the beginning and the corresponding values in the dimension statement. Type up the program and execute it.

```
 10    REM Simultaneous solution of linear equations
 11    REM By Gauss elimination, Apr 14, 81
 12    REM identifiers
 14    REM      C1      COEF         solution vector
 15    REM      E1%     ERMES%       error flag
 16    REM      M1%     MAX%         maximum length
 17    REM      N1%     NROW%        number of rows
 18    REM      N2%     NCOL%        number of columns
 19    REM end of identifiers
 30    REM
 40    A$ = " ##.###^^^^ "
 50    B$ = "  =  ##.###^^^^"
 60    C$ = "  ##.#####"
 70    M1% = 8
 80    DIM Z(8), A(8,8), C1(8), W(8), B(8,8)
 90    PRINT "Simultaneous solution by Gauss elimination"
100    GOSUB 500 : REM input subroutine
110    GOSUB 5000 : REM Gauss elimination
120    REM
130    IF (N1% > 5) THEN 210
140    PRINT "          Matrix    Constants"
150    FOR I% = 1 TO N1%
160       FOR J% = 1 TO N2%
170          PRINT USING A$; A(I%,J%);
180       NEXT J%
190       PRINT USING B$; Z(I%)
200    NEXT I%
210    PRINT
220    IF (E1% = 1) THEN 290
230    PRINT "          Solution"
240    PRINT
250    FOR I% = 1 TO N2%
260       PRINT USING C$; C1(I%);
270    NEXT I%
```

Figure 4.4: Solution of Simultaneous Equations by Gauss Elimination

```
280    PRINT
290    GOTO 100 : REM next set of equations
500    REM
510    REM input the data
520    REM
530    PRINT
540    INPUT " How  many  equations"; N1%
550    IF (N1% > M1%) THEN 540
560    IF (N1% < 2) THEN 9999
570    N2% = N1%
580    FOR I% = 1 TO N1%
590      PRINT "Equation"; I%
600      FOR J% = 1 TO N2%
610        PRINT J%; "   ";
620        INPUT A(I%,J%)
630      NEXT J%
640      INPUT "   C "; Z(I%)
650    NEXT I%
660    RETURN : REM from input subroutine
5000   REM Simultaneous solution by Gauss elimination
5010   REM Apr 14, 81
5011   REM identifiers
5013   REM      A       A           coefficient matrix
5014   REM      B       B           work array
5015   REM      B1      BIG
5016   REM      C1      COEF        solution vector
5017   REM      E1%     ERMES%      error flag
5018   REM      H1      HOLD
5019   REM      N2%     NCOL%       number of columns
5020   REM      S6      SUM
5021   REM      Z       Z           constant vector
5022   REM end of identifiers
5070   REM
```

Figure 4.4:
Solution of Simultaneous Equations by Gauss Elimination (cont.)

```
5080    FOR I% = 1 TO N2%
5090      FOR J% = 1 TO N2%
5100        B(I%,J%) = A(I%,J%)
5110      NEXT J%
5120      W(I%) = Z(I%)
5130    NEXT I%
5140    E1% = 0
5150    FOR I% = 1 TO N2% − 1
5160      B1 = ABS(B(I%,I%))
5170      L% = I%
5180      I1% = I% + 1
5190      FOR J% = I1% TO N2%
5200        IF (ABS(B(J%,I%)) < B1) THEN 5230
5210        B1 = ABS (B(J%,I%))
5220        L% = J%
5230      NEXT J%
5240      IF (B1 = 0) THEN 5530
5250      IF (L% = I%) THEN 5340
5260      FOR J% = 1 TO N2%
5270        H1 = B(L%,J%)
5280        B(L%,J%) = B(I%,J%)
5290        B(I%,J%) = H1
5300      NEXT J%
5310      H1 = W(L%)
5320      W(L%) = W(I%)
5330      W(I%) = H1
5340      FOR J% = I1% TO N2%
5350        T = B(J%,I%) / B(I%,I%)
5360        FOR K% = I1% TO N2%
5370          B(J%,K%) = B(J%,K%) − T * B(I%,K%)
5380        NEXT K%
5390        W(J%) = W(J%) − T * W(I%)
5400      NEXT J%
5410    NEXT I%
```

Figure 4.4:
Solution of Simultaneous Equations by Gauss Elimination (cont.)

```
5420    IF (B(N2%,N2%) = 0) THEN 5530
5430    C1(N2%) = W(N2%) / B(N2%,N2%)
5440    REM back substitution
5450    FOR I% = N2% − 1 TO 1 STEP −1
5460        S6 = 0
5470        FOR J% = I% + 1 TO N2%
5480            S6 = S6 + B(I%,J%) * C1(J%)
5490        NEXT J%
5500        C1(I%) = (W(I%) − S6) / B(I%,I%)
5510    NEXT I%
5520    RETURN : REM normal return
5530    E1% = 1
5540    PRINT "ERROR—Matrix singular"
5550    RETURN : REM from Gauss—elimination subroutine
9999    END
```

Figure 4.4:
Solution of Simultaneous Equations by Gauss Elimination (cont.)

Running the Program

This program is similar to the previous one, but it begins by asking for the number of equations. Answer this question with a number from 2 to 8, then give a carriage return. The coefficients for each equation and the corresponding constant vectors are entered in turn. Press the carriage return after each number is entered. Enter the values for the set of equations we considered earlier in this chapter and verify that the solution vector is [3 2 1]. The results should look like Figure 4.5.

At the completion of the task, the program begins again. This time find the simultaneous solution to the following three equations and write down the answer for later reference.

$$\begin{bmatrix} 1 & 1 & 1 \\ 2 & 1 & -1 \\ 3 & 1 & -3 \end{bmatrix} \quad \begin{bmatrix} 6 \\ 1 \\ -4 \end{bmatrix}$$

We will come back to this set at the end of the next section. For the third task, enter two lines that are exactly the same, for example:

$$\begin{bmatrix} 1 & 1 & 1 \\ 1 & 1 & 1 \\ 2 & 1 & -1 \end{bmatrix} \quad \begin{bmatrix} 6 \\ 6 \\ 1 \end{bmatrix}$$

```
Simultaneous solution by Gauss elimination

How many equations? 3
Equation 1
1     ? 13
2     ? -8
3     ? -3
C ? 20
Equation 2
1     ? -8
2     ? 10
3     ? -1
C ? -5
Equation 3
1     ? -3
2     ? -1
3     ? 11
C ? 0

Matrix    Constants
 1.300E+01   -8.000E+00   -3.000E+00   =   2.000E+01
-8.000E+00    1.000E+01   -1.000E+00   =  -5.000E+00
-3.000E+00   -1.000E+00    1.100E+01   =   0.000E+00
Solution
3.00000    2.00000    1.00000
```

Figure 4.5: Input/Output Dialogue: Solving Simultaneous Equations By Gauss Elimination

The program should print an error message indicating that the matrix is singular. The program can be aborted by entering a value less than 2 for the number of equations.

SOLUTION BY GAUSS-JORDAN ELIMINATION

A variation of the Gauss elimination method is known as the Gauss-Jordan method. The approach shares most of the advantages and disadvantages of the Gauss elimination technique. Execution time is a third-order function of the matrix size and there are many multiplication, division, and subtraction operations that contribute to loss of accuracy. Furthermore, the Gauss-Jordan algorithm is more complicated than the one for Gauss elimination. Nevertheless, the Gauss-Jordan technique will generally be the most useful of all. In fact, we will use it in later chapters of this book. Its usefulness lies in the fact that the inverse of the coefficient matrix is readily obtained along with the solution vector. Let us outline the steps of this method.

Details of the Gauss-Jordan Method

In the Gauss-Jordan method, the elements of the major diagonal are converted to unity as they are in the Gauss method. But now the elements both above and below the major diagonal are converted to zeros. Thus, the coefficient matrix is converted to a unit matrix. The resulting constant vector then becomes the solution vector. For the Gauss-Jordan solution of the electrical circuit given in Figure 4.1, the final set of values becomes:

$$\begin{bmatrix} 1 & 0 & 0 \\ 0 & 1 & 0 \\ 0 & 0 & 1 \end{bmatrix} \quad \begin{bmatrix} 3 \\ 2 \\ 1 \end{bmatrix}$$

This corresponds to the three equations:

$$\begin{aligned} x &= 3 \\ y &= 2 \\ z &= 1 \end{aligned}$$

for which the solution is:

$$x = 3, \quad y = 2, \quad z = 1$$

Suppose that a unit matrix is initially placed to the right of the original set of equations. Then, if all of the operations are performed on this matrix as they are performed on the other elements, this unit matrix will be converted into the inverse of the original coefficient matrix at the conclusion of the calculation. That is, the set:

$$\begin{bmatrix} 13 & -8 & -3 \\ -8 & 10 & -1 \\ -3 & -1 & 11 \end{bmatrix} \quad \begin{bmatrix} 20 \\ -5 \\ 0 \end{bmatrix} \quad \begin{bmatrix} 1 & 0 & 0 \\ 0 & 1 & 0 \\ 0 & 0 & 1 \end{bmatrix}$$

will be transformed into:

$$\begin{bmatrix} 1 & 0 & 0 \\ 0 & 1 & 0 \\ 0 & 0 & 1 \end{bmatrix} \quad \begin{bmatrix} 3 \\ 2 \\ 1 \end{bmatrix} \quad \begin{bmatrix} 0.19 & 0.16 & 0.07 \\ 0.16 & 0.23 & 0.06 \\ 0.07 & 0.06 & 0.11 \end{bmatrix}$$

In this case, the matrix on the right is the inverse of the original coefficient matrix.

It is not actually necessary to use two separate matrices for these operations, however. The inverse matrix can be physically generated in the same space occupied by the original coefficient matrix. At the conclusion of the operation, the inverse matrix will then be returned in place of the coefficient matrix. And, of course, the solution vector will appear in place of the original constant vector. Thus, if the coefficient matrix and constant vector:

$$\begin{bmatrix} 13 & -8 & -3 \\ -8 & 10 & -1 \\ -3 & -1 & 11 \end{bmatrix} \qquad \begin{bmatrix} 20 \\ -5 \\ 0 \end{bmatrix}$$

are passed to the Gauss-Jordan routine, the matrix inverse and solution vector can be returned in the same array space:

$$\begin{bmatrix} 0.19 & 0.16 & 0.07 \\ 0.16 & 0.23 & 0.06 \\ 0.07 & 0.06 & 0.11 \end{bmatrix} \qquad \begin{bmatrix} 3 \\ 2 \\ 1 \end{bmatrix}$$

We will now present a BASIC program that implements this rather complex algorithm.

BASIC PROGRAM: GAUSS-JORDAN ELIMINATION

The program shown in Figure 4.6 will solve simultaneous linear equations by the Gauss-Jordan elimination method. The main program calls the Gauss-Jordan subroutine at line 5000. With this version, the original matrix A is duplicated in matrix B at the beginning. The inverse of the coefficient matrix is returned in matrix B in place of the original data. The data vector, Z, is also isolated from the solution vector, C1. The matrix inverse and the determinant of the coefficient matrix can also be displayed. Change the value of 15% at line 5100 to a value of 1 to select this option.

```
10   REM Simultaneous solution of linear equations
11   REM by Gauss — Jordan elimination, Apr 20, 81
12   REM identifiers
14   REM      C1      COEF        solution vector
15   REM      E1%     ERMES%      error flag
16   REM      I2%     INDEX%      work matrix
17   REM      M1%     MAX%        maximum length
18   REM      N1%     NROW%       number of rows
19   REM      N2%     NCOL%       number of columns
20   REM end of identifiers
30   REM
```

Figure 4.6: Solution of Simultaneous Equations by Gauss-Jordan Elimination

```
 40    A$ = '' ##.###^^^ ''
 50    B$ = ''  =  ##.###^^^''
 60    C$ = ''   ##.#####''
 70    M1% = 8
 80    DIM Z(8), A(8,8), C1(8), W(8,1), B(8,8), I2%(8,3)
 90    REM
100    PRINT "Simultaneous solution by Gauss—Jordan elimination"
110    GOSUB 500 : REM get the data
120    GOSUB 5000 : REM Gauss—Jordan subroutine
130    REM
140    IF (N1% > 5) THEN 220
150    PRINT ''           Matrix     Constants''
160    FOR I% = 1 TO N1%
170       FOR J% = 1 TO N2%
180          PRINT USING A$; A(I%,J%);
190       NEXT J%
200       PRINT USING B$; Z(I%)
210    NEXT I%
220    PRINT
230    IF (E1% = 1) THEN 300
240    PRINT ''           Solution''
250    PRINT
260    FOR I% = 1 TO N2%
270       PRINT USING C$; C1(I%);
280    NEXT I%
290    PRINT
300    GOTO 110 : REM next set of equations
500    REM
510    REM input the data
520    REM
530    PRINT
540    INPUT '' How  many  equations''; N1%
550    IF (N1% > M1%) THEN 540
560    IF (N1% < 2) THEN 9999
```

Figure 4.6: Solution of Simultaneous Equations by Gauss-Jordan Elimination (cont.)

```
570   N2% = N1%
580   FOR I% = 1 TO N1%
590     PRINT "Equation"; I%
600     FOR J% = 1 TO N1%·
610       PRINT J%; "   ";
620       INPUT A(I%,J%)
630     NEXT J%
640     INPUT "   C "; Z(I%)
650   NEXT I%
660   RETURN : REM from·input subroutine
5000  REM Gauss—Jordan matrix inversion and solution
5010  REM Apr 20, 81
5011  REM identifiers
5013  REM     A        A          coefficient matrix
5014  REM     B        B          work matrix
5015  REM     B1       BIG        biggest value
5016  REM     C1       COEF       solution vector
5017  REM     D3       DETERM     determinant
5018  REM     E1%      ERMES%     error flag
5019  REM     H1       HOLD       work variable
5020  REM     I2%      INDEX%     work matrix
5021  REM     I3%      IROW%      row index
5022  REM     I4%      ICOL%      column index
5023  REM     I5%      INVRS%     print—inverse flag
5024  REM     N2%      NCOL%      number of columns
5025  REM     N3%      NVEC%      number of constant vectors
5026  REM     P1       PIVOT      pivot index
5027  REM     W        W          solution matrix
5028  REM     Z        Z          constant vector
5029  REM end of identifiers
5080  REM
5090  E1% = 0 : REM becomes 1 for singular matrix
5100  I5% = 1 : REM print inverse matrix if zero
5110  N3% = 1 : REM number of constant vectors
```

Figure 4.6:
Solution of Simultaneous Equations by Gauss-Jordan Elimination (cont.)

```
5120    FOR I% = 1 TO N2%
5130      FOR J% = 1 TO N2%
5140        B(I%,J%) = A(I%,J%)
5150      NEXT J%
5160      W(I%,1) = Z(I%)
5170      I2%(I%,3) = 0
5180    NEXT I%
5190    D3 = 1
5200    FOR I% = 1 TO N2%
5210      REM
5220      REM search for largest (pivot) element
5230      REM
5240      B1 = 0
5250      FOR J% = 1 TO N2%
5260        IF (I2%(J%,3) = 1) THEN 5350
5270        FOR K% = 1 TO N2%
5280          IF (I2%(K%,3) > 1) THEN 6120
5290          IF (I2%(K%,3) = 1) THEN 5340
5300          IF (B1 >= ABS(B(J%,K%))) THEN 5340
5310          I3% = J%
5320          I4% = K%
5330          B1 = ABS(B(J%,K%))
5340        NEXT K%
5350      NEXT J%
5360      I2%(I4%,3) = I2%(I4%,3) + 1
5370      I2%(I%,1) = I3%
5380      I2%(I%,2) = I4%
5390      REM interchange rows to put pivot on diagonal
5400      IF (I3% = I4%) THEN 5540
5410      D3 = -D3
5420      FOR L% = 1 TO N2%
5430        H1 = B(I3%,L%)
5440        B(I3%,L%) = B(I4%,L%)
5450        B(I4%,L%) = H1
```

Figure 4.6:
Solution of Simultaneous Equations by Gauss-Jordan Elimination (cont.)

```
5460      NEXT L%
5470      IF (N3% < 1) THEN 5540
5480      FOR L% = 1 TO N3%
5490          H1 = W(I3%,L%)
5500          W(I3%,L%) = W(I4%,L%)
5510          W(I4%,L%) = H1
5520      NEXT L%
5530      REM divide pivot row by pivot element
5540      P1 = B(I4%, I4%)
5550      D3 = D3 * P1
5560      B(I4%,I4%) = 1
5570      FOR L% =1 TO N2%
5580          B(I4%,L%) = B(I4%,L%) / P1
5590      NEXT L%
5600      IF (N3% < 1) THEN 5660
5610      FOR L% = 1 TO N3%
5620          W(I4%,L%) = W(I4%,L%) / P1
5630      NEXT L%
5640      REM
5650      REM reduce nonpivot rows
5660      FOR L1% = 1 TO N2%
5670          IF (L1% = I4%) THEN 5770
5680          T = B(L1%, I4%)
5690          B(L1%,I4%) = 0
5700          FOR L% = 1 TO N2%
5710              B(L1%,L%) = B(L1%,L%) − B(I4%,L%) * T
5720          NEXT L%
5730          IF (N3% < 1) THEN 5770
5740          FOR L% = 1 TO N3%
5750              W(L1%,L%) = W(L1%,L%) − W(I4%,L%) * T
5760          NEXT L%
5770      NEXT L1%
5780      NEXT I%
5790      REM
5800      REM interchange columns
```

Figure 4.6:
Solution of Simultaneous Equations by Gauss-Jordan Elimination (cont.)

```
5810    REM
5820    FOR I% = 1 TO N2%
5830        L% = N2% − I% + 1
5840        IF (I2%(L%,1) = I2%(L%,2)) THEN 5920
5850        I3% = I2%(L%,1)
5860        I4% = I2%(L%,2)
5870        FOR K% = 1 TO N2%
5880            H1 = B(K%, I3%)
5890            B(K%,I3%) = B(K%,I4%)
5900            B(K%,I4%) = H1
5910        NEXT K%
5920    NEXT I%
5930    FOR K% = 1 TO N2%
5940        IF (I2%(K%,3) <> 1) THEN 6120
5950    NEXT K%
5960    E1% = 0
5970    FOR I% = 1 TO N2%
5980        C1(I%) = W(I%,1)
5990    NEXT I%
6000    IF (I5% = 1) THEN 6140
6010    PRINT
6020    PRINT "    Matrix inverse"
6030    FOR I% = 1 TO N2%
6040        FOR J% = 1 TO N2%
6050            PRINT USING A$; B(I%,J%);
6060        NEXT J%
6070        PRINT
6080    NEXT I%
6090    PRINT
6100    PRINT "Determinant = "; D3
6110    RETURN : REM if inverse is printed
6120    E1% = 1
6130    PRINT "ERROR—matrix singular "
6140    RETURN : REM from Gauss—Jordan subroutine
9999    END
```

Figure 4.6:
Solution of Simultaneous Equations by Gauss-Jordan Elimination (cont.)

Type up the program and execute it. Try out the electrical circuit equations and verify that the solution is [3 2 1]. Then try the set:

$$\begin{bmatrix} 1 & 1 & 1 \\ 2 & 1 & -1 \\ 3 & 1 & -3 \end{bmatrix} \qquad \begin{bmatrix} 6 \\ 1 \\ -4 \end{bmatrix}$$

Notice that the answer, in this case, is not the same as the one found by the Gauss method. Furthermore, the set:

$$x = 1, \quad y = 2, \quad z = 3$$

is also a solution. How can there be more than one solution to a set of linear equations? The answer is that one of the equations is a linear combination of the other two. The third equation, in this case, is equal to twice the second equation minus the first equation. Thus, the coefficient matrix is singular. Actually, if the above three equations are given to the Cramer's rule program at the beginning of this chapter, the singularity will be readily found.

Unfortunately, this kind of linear dependence is difficult to find. The determinant of the coefficient matrix may be small, but it does not equal zero because of roundoff errors that accumulate during the elimination process. Fortunately, different algorithms for solving simultaneous equations will usually give different answers in this case. Consequently, if there is any question of linear dependence, the equations should be solved by at least two different methods.

We have seen how the Gauss-Jordan elimination program works for one coefficient matrix and its corresponding constant vector. Now we will explore ways of using and refining this program to deal conveniently with *multiple* constant vectors. To illustrate this situation we will return to the electrical circuit example that we set up earlier in this chapter. We will also continue our discussion of inverse coefficient matrices and their use in solving simultaneous equations.

MULTIPLE CONSTANT VECTORS AND MATRIX INVERSION

In the previous chapter, we mentioned that the solution to a set of linear equations could be obtained by multiplying the inverse of the coefficient matrix by the constant vector. For example, if the coefficient matrix is A and the constant vector is $\mathbf{y}$, then the solution vector $\mathbf{b}$ is:

$$\mathbf{b} = A^{-1} \mathbf{y}$$

The methods developed in this chapter (Cramer's rule, Gauss elimination, and Gauss-Jordan elimination) obtain the solution to a set of linear equations by direct methods, however. The inverse of the coefficient matrix is not utilized.

Sometimes we need to solve several sets of simultaneous equations that all have the same coefficient matrix but different constant vectors. In this case, we can invert the coefficient matrix. Then, the separate solution vectors can be obtained from the product of the inverted matrix and each constant vector. Even in this case, however, it will generally be faster to perform a Gauss-Jordan elimination on the matrix and all of the constant vectors simultaneously. In fact, the Gauss-Jordan routine given in the previous section is actually programmed for multiple constant vectors.

Consider, for example, the electrical circuit shown in Figure 4.1. Suppose that we would like to determine the loop currents for three different circuit configurations:

1. The original circuit.

2. The circuit with the 5-volt source reversed.

3. The circuit with both voltage sources reversed.

These three different configurations correspond to the following three sets of equations:

$$\begin{bmatrix} 13 & -8 & -3 \\ -8 & 10 & -1 \\ -3 & -1 & 11 \end{bmatrix} \quad \begin{bmatrix} 20 \\ -5 \\ 0 \end{bmatrix}$$

$$\begin{bmatrix} 13 & -8 & -3 \\ -8 & 10 & -1 \\ -3 & -1 & 11 \end{bmatrix} \quad \begin{bmatrix} 20 \\ 5 \\ 0 \end{bmatrix}$$

$$\begin{bmatrix} 13 & -8 & -3 \\ -8 & 10 & -1 \\ -3 & -1 & 11 \end{bmatrix} \quad \begin{bmatrix} -20 \\ 5 \\ 0 \end{bmatrix}$$

All three sets of equations can be solved separately using one of the preceding techniques. In this case, however, it is more efficient to solve all three conditions at once by using the Gauss-Jordan method. The coefficient matrix contains the values common to the three different circuits. However, the constant vector now becomes a *constant matrix*. Each column of the constant matrix represents a different circuit configuration and will produce the corresponding solution. The actual matrices given to the Gauss-Jordan routine are:

$$\begin{bmatrix} 13 & -8 & -3 \\ -8 & 10 & -1 \\ -3 & -1 & 11 \end{bmatrix} \qquad \begin{bmatrix} 20 & 20 & -20 \\ -5 & 5 & 5 \\ 0 & 0 & 0 \end{bmatrix}$$

The Gauss-Jordan routine returns the following values:

$$\begin{bmatrix} 0.19 & 0.16 & 0.07 \\ 0.16 & 0.23 & 0.06 \\ 0.07 & 0.06 & 0.11 \end{bmatrix} \qquad \begin{bmatrix} 3 & 4.58 & -3 \\ 2 & 4.33 & -2 \\ 1 & 1.64 & -1 \end{bmatrix}$$

The left matrix, which originally contained the coefficients, now holds the inverse of the coefficient matrix. The right matrix, which initially contained the constant vectors, now contains the corresponding solution vectors. Thus, for the three separate circuits, the answers are:

		Solution	
Circuit	I_1	I_2	I_3
1	3	2	1
2	4.58	4.33	1.64
3	-3	-2	-1

In the previous version of the Gauss-Jordan program, we copied the original coefficient matrix A into a work array B at the beginning of the Gauss-Jordan subroutine. Also, the incoming constant vector Z was copied into the first column of the constant matrix W at this time. At the end of the Gauss-Jordan subroutine, the solution vector in the first column of matrix W was copied into the solution vector C1. Now we will look at a refined version of this program.

BASIC PROGRAM: GAUSS-JORDAN ELIMINATION, VERSION TWO

An alternate version of the Gauss-Jordan program is given in Figure 4.8. Some of the arrays have been changed. As before, the inverse of the coefficient matrix is returned in array B, which originally held the coefficient matrix. But the solution matrix is now returned in the array that originally contained the matrix of constant vectors. In addition, the determinant of the coefficient matrix is available as a separate parameter.

Running the Program

Type up the program and execute it. The user will be asked for the number of equations, as before. Next will be a question about the number of constant vectors. If this question is answered with the value of unity, then the program will behave exactly as before. However, the solution vector is now printed vertically rather than horizontally.

With this version, the number of constant vectors may be greater than one. In this case, the constants for each equation are entered with the corresponding coefficients. For example, if the above three electrical circuits are collectively solved with this program, then the data will be entered as:

$$13 \quad -8 \quad -3 \quad 20 \quad 20 \quad -20$$

Solve all three of the above electric circuits at the same time and verify that the correct answers are obtained.

This new version has another option. The number of constant vectors may be zero. In this case, only a matrix of coefficients is entered. The program will then print the inverse of the coefficient matrix. The determinant of the coefficient matrix will also be displayed.

Enter the coefficient matrix we previously had trouble with:

$$\begin{bmatrix} 1 & 1 & 1 \\ 2 & 1 & -1 \\ 3 & 1 & -3 \end{bmatrix}$$

This matrix is singular and so the determinant is zero. However, this Gauss-Jordan program may return a small, but nonzero, value because of roundoff error. The result for this singular matrix will look like Figure 4.7. (Enter the value of zero to abort the program.)

```
Simultaneous solution by Gauss-Jordan elimination

How many equations? 3
How many constant vectors? 0
Equation 1
1    ?  1
2    ?  1
3    ?  1
Equation 2
1    ?  2
2    ?  1
3    ?  -1
Equation 3
1    ?  3
2    ?  1
3    ?  -3
```

Figure 4.7: Input/Output Dialogue: Solving Simultaneous Equations by Gauss-Jordan Elimination

```
Matrix inverse
-1.118E+07    2.237E+07   -1.118E+07
 1.678E+07   -3.355E+07    1.678E+07
-5.592E+06    1.118E+07   -5.592E+06

Determinant =  1.78814E-07
Matrix
1.000E+00    1.000E+00    1.000E+00
2.000E+00    1.000E+00   -1.000E+00
3.000E+00    1.000E+00   -3.000E+00

How many equations? 0
```

Figure 4.7: Input/Output Dialogue:
Solving Simultaneous Equations by Gauss-Jordan Elimination (cont.)

Notice that in this example, the elements of the inverted matrix are many orders of magnitude larger than the original elements. This is an indication of ill conditioning, a subject we will consider in the next section.

```
10    REM Simultaneous solution with multiple constants
11    REM by Gauss — Jordan elimination, Apr 20, 81
12    REM identifiers
14    REM      E1%    ERMES%    error flag
15    REM      I2%    INDEX%    work matrix
16    REM      I5%    INVRS%    inverse flag
17    REM      M1%    MAX%      maximum length
18    REM      N1%    NROW%     number of rows
19    REM      N2%    NCOL%     number of columns
20    REM      N3%    NVEC%     number of constant vectors
21    REM end of identifiers
40    REM
50    A$ = " ##.###^^^ "
60    B$ = " = ##.###^^^"
70    C$ = " ##.#####"
80    M1% = 8
90    DIM A(8,8), Z(8,8), W(8,8), B(8,8), I2%(8,3)
100   REM
110   PRINT "Simultaneous solution by Gauss — Jordan elimination"
```

Figure 4.8: Solution of Simultaneous Equations
and Matrix Inversion by the Gauss-Jordan Method (Multiple constant vectors may be entered.)

```
120    GOSUB 500 : REM get the data
130    GOSUB 5000 : REM Gauss—Jordan subroutine
140    REM
150    IF (N1% > 5) THEN 290
160    PRINT "          Matrix";
170    IF (N3% > 0) THEN PRINT "        Constants";
180    PRINT
190    FOR I% = 1 TO N1%
200      FOR J% = 1 TO N2%
210        PRINT USING A$; A(I%,J%);
220      NEXT J%
230      IF (N3% = 0) THEN 270
240      FOR J% = 1 TO N3% : REM constant vectors
250        PRINT USING B$; Z(I%,J%);
260      NEXT J%
270      PRINT
280    NEXT I%
290    PRINT
300    IF (E1% = 1 OR N3% = 0) THEN 400
310    PRINT "          Solution"
320    PRINT
330    FOR I% = 1 TO N2%
340      FOR J% = 1 TO N3%
350        PRINT USING C$; W(I%,J%);
360      NEXT J%
370      PRINT
380    NEXT I%
390    PRINT
400    GOTO 120 : REM next set of equations
500    REM
510    REM input the data
520    REM
530    I5% = 1
540    PRINT
```

Figure 4.8: Solution of Simultaneous Equations and Matrix Inversion by the Gauss-Jordan Method (Multiple constant vectors may be entered.) (cont.)

```
550    INPUT " How many equations"; N1%
560    IF (N1% > M1%) THEN 550
570    IF (N1% < 2) THEN 9999
580    N2% = N1%
590    INPUT " How many constant vectors"; N3%
600    IF (N3% = 0) THEN I5% = 0 : REM print inverse
610    FOR I% = 1 TO N1%
620       PRINT "Equation"; I%
630       FOR J% = 1 TO N2%
640          PRINT J%; "   ";
650          INPUT A(I%,J%) : REM coefficients
660       NEXT J%
670       IF (N3% < 1) THEN 710
680       FOR J% = 1 TO N3%
690          INPUT "  C "; Z(I%,J%) : REM constant vector
700       NEXT J%
710    NEXT I%
720    RETURN : REM from input routine
5000   REM Gauss — Jordan matrix inversion and solution
5010   REM for multiple constant vectors, Apr 20, 81
5011   REM identifiers
5013   REM      A        A          coefficient matrix
5014   REM      B        B          work matrix
5015   REM      B1       BIG        largest element
5016   REM      D3       DETERM     determinant
5017   REM      E1%      ERMES%     error flag
5018   REM      H1       HOLD       work variable
5019   REM      I2%      INDEX%     work matrix
5020   REM      I3%      IROW%      row index
5021   REM      I4%      ICOL%      column index
5022   REM      I5%      INVRS%     print — inverse flag
5023   REM      N2%      NCOL%      number of columns
5024   REM      N3%      NVEC%      number of constant vectors
5025   REM      P1       PIVOT      pivot index
```

Figure 4.8: Solution of Simultaneous Equations and Matrix Inversion by the Gauss-Jordan Method (Multiple constant vectors may be entered.) (cont.)

```
5026   REM      W       W              solution matrix
5027   REM      Z       Z              constant vector
5028   REM end of identifiers
5090   REM
5100   E1% = 0 : REM becomes 1 for singular matrix
5110   REM I5% = 1 : REM print inverse if zero
5120   REM
5130   REM N3% = 1
5140   FOR I% = 1 TO N2%
5150      FOR J% = 1 TO N2%
5160         B(I%,J%) = A(I%,J%)
5170      NEXT J%
5180      FOR J% = 1 TO N3%
5190         W(I%,J%) = Z(I%,J%)
5200      NEXT J%
5210      I2%(I%,3) = 0
5220   NEXT I%
5230   D3 = 1
5240   FOR I% = 1 TO N2%
5250      REM
5260      REM search for largest (pivot) element
5270      REM
5280      B1 = 0
5290      FOR J% = 1 TO N2%
5300         IF (I2%(J%,3) = 1) THEN 5390
5310         FOR K% = 1 TO N2%
5320            IF (I2%(K%,3) > 1) THEN 6130
5330            IF (I2%(K%,3) = 1) THEN 5380
5340            IF (B1 >= ABS(B(J%,K%))) THEN 5380
5350            I3% = J%
5360            I4% = K%
5370            B1 = ABS(B(J%,K%))
5380         NEXT K%
5390      NEXT J%
```

Figure 4.8: Solution of Simultaneous Equations and Matrix Inversion by the Gauss-Jordan Method (Multiple constant vectors may be entered.) (cont.)

```
5400    I2%(I4%,3) = I2%(I4%,3) + 1
5410    I2%(I1%,1) = I3%
5420    I2%(I1%,2) = I4%
5430    REM interchange rows to put pivot on diagonal
5440    IF (I3% = I4%) THEN 5580
5450    D3 = -D3
5460    FOR L% = 1 TO N2%
5470        H1 = B(I3%,L%)
5480        B(I3%,L%) = B(I4%,L%)
5490        B(I4%,L%) = H1
5500    NEXT L%
5510    IF (N3% < 1) THEN 5580
5520    FOR L% = 1 TO N3%
5530        H1 = W(I3%,L%)
5540        W(I3%,L%) = W(I4%,L%)
5550        W(I4%,L%) = H1
5560    NEXT L%
5570    REM divide pivot row by pivot element
5580    P1 = B(I4%, I4%)
5590    D3 = D3 * P1
5600    B(I4%,I4%) = 1
5610    FOR L% = 1 TO N2%
5620        B(I4%,L%) = B(I4%,L%) / P1
5630    NEXT L%
5640    IF (N3% < 1) THEN 5700
5650    FOR L% = 1 TO N3%
5660        W(I4%,L%) = W(I4%,L%) / P1
5670    NEXT L%
5680    REM
5690    REM reduce nonpivot rows
5700    FOR L1% = 1 TO N2%
5710        IF (L1% = I4%) THEN 5810
5720        T = B(L1%, I4%)
5730        B(L1%,I4%) = 0
```

Figure 4.8: Solution of Simultaneous Equations and Matrix Inversion by the Gauss-Jordan Method (Multiple constant vectors may be entered.) (cont.)

```
5740          FOR L% = 1 TO N2%
5750             B(L1%,L%) = B(L1%,L%) − B(I4%,L%) ∗ T
5760          NEXT L%
5770          IF (N3% < 1) THEN 5810
5780          FOR L% = 1 TO N3%
5790             W(L1%,L%) = W(L1%,L%) − W(I4%,L%) ∗ T
5800          NEXT L%
5810        NEXT L1%
5820      NEXT I%
5830      REM
5840      REM interchange columns
5850      REM
5860      FOR I% = 1 TO N2%
5870        L% = N2% − I% + 1
5880        IF (I2%(L%,1) = I2%(L%,2)) THEN 5960
5890        I3% = I2%(L%,1)
5900        I4% = I2%(L%,2)
5910        FOR K% = 1 TO N2%
5920           H1 = B(K%, I3%)
5930           B(K%,I3%) = B(K%,I4%)
5940           B(K%,I4%) = H1
5950        NEXT K%
5960      NEXT I%
5970      FOR K% = 1 TO N2%
5980        IF (I2%(K%,3) <> 1) THEN 6130
5990      NEXT K%
6000      E1% = 0
6010      IF (I5% = 1) THEN 6150
6020      PRINT
6030      PRINT "    Matrix inverse"
6040      FOR I% = 1 TO N2%
6050        FOR J% = 1 TO N2%
6060           PRINT USING A$; B(I%,J%);
6070        NEXT J%
```

Figure 4.8: Solution of Simultaneous Equations and Matrix Inversion by the Gauss-Jordan Method (Multiple constant vectors may be entered.) (cont.)

```
6080      PRINT
6090      NEXT I%
6100      PRINT
6110      PRINT "Determinant = "; D3
6120      RETURN : REM if inverse is printed
6130      E1% = 1
6140      PRINT "ERROR—matrix singular "
6150      RETURN : REM from Gauss—Jordan subroutine
9999      END
```

Figure 4.8: Solution of Simultaneous Equations and Matrix Inversion by the Gauss-Jordan Method (Multiple constant vectors may be entered.) (cont.)

In the following section we will discuss the Hilbert matrix as an example of ill conditioning. To explore such matrices we will write a program that uses our first version of the Gauss-Jordan method (Figure 4.6). We will then use the program to solve a series of progressively more problematic Hilbert matrices.

ILL-CONDITIONED EQUATIONS

A singular matrix has a determinant of zero. The corresponding set of equations has either no solutions or many solutions. An *ill-conditioned matrix*, by comparison, is one that is *nearly* singular. It produces incorrect or inaccurate answers. A small change in the input data can cause great changes in the answer. A two-dimensional analogue of ill conditioning occurs with two nearly parallel lines. The point of intersection is the desired solution; but the closer the two lines come to being parallel, the more difficult it is to determine their actual point of intersection.

Various tests for ill conditioning have been suggested. One of these is to compare the values of the inverted matrix to those of the original matrix. If there are differences of several orders of magnitude, then it is likely that ill conditioning is present. Another test is to take the inverse of the inverse of the coefficient matrix and compare the result to the original matrix. The two should, of course, be the same. This technique will test the inversion algorithm and the computer arithmetic at the same time. Yet another test is to compare the magnitudes of values along the major diagonal. They should not be too far apart.

The Hilbert matrix is an example of ill conditioning. This symmetric matrix begins with unity in the upper-left corner. The remaining values get smaller and smaller as we go down a column or across a row, according to the pattern:

$$\begin{bmatrix} 1 & \frac{1}{2} & \frac{1}{3} & \frac{1}{4} & \cdots\cdots & 1/n \\ \frac{1}{2} & \frac{1}{3} & \frac{1}{4} & \frac{1}{5} & \cdots & 1/(n+1) \\ \frac{1}{3} & \frac{1}{4} & \frac{1}{5} & \frac{1}{6} & \cdots & 1/(n+2) \\ \frac{1}{4} & \frac{1}{5} & \frac{1}{6} & \frac{1}{7} & \cdots & 1/(n+3) \\ \cdots & & & \cdots & & \cdots \\ 1/n & \cdots\cdots\cdots\cdots\cdots & & & & 1/(2n-1) \end{bmatrix}$$

The Hilbert matrix can be used to produce a set of ill-conditioned equations. Consider, for example:

$$x_1 + \frac{x_2}{2} + \frac{x_3}{3} + \frac{x_4}{4} + \frac{x_5}{5} = 1 + \frac{1}{2} + \frac{1}{3} + \frac{1}{4} + \frac{1}{5}$$

$$\frac{x_1}{2} + \frac{x_2}{3} + \frac{x_3}{4} + \frac{x_4}{5} + \frac{x_5}{6} = \frac{1}{2} + \frac{1}{3} + \frac{1}{4} + \frac{1}{5} + \frac{1}{6}$$

$$\frac{x_1}{3} + \frac{x_2}{4} + \frac{x_3}{5} + \frac{x_4}{6} + \frac{x_5}{7} = \frac{1}{3} + \frac{1}{4} + \frac{1}{5} + \frac{1}{6} + \frac{1}{7}$$

$$\frac{x_1}{4} + \frac{x_2}{5} + \frac{x_3}{6} + \frac{x_4}{7} + \frac{x_5}{8} = \frac{1}{4} + \frac{1}{5} + \frac{1}{6} + \frac{1}{7} + \frac{1}{8}$$

$$\frac{x_1}{5} + \frac{x_2}{6} + \frac{x_3}{7} + \frac{x_4}{8} + \frac{x_5}{9} = \frac{1}{5} + \frac{1}{6} + \frac{1}{7} + \frac{1}{8} + \frac{1}{9}$$

First of all, the fractions 1/3, 1/6 and 1/7 cannot be represented exactly. Consequently, there will be roundoff errors at the very beginning of the problem. Secondly, the inverse of the coefficient matrix is exactly:

$$\begin{bmatrix} 25 & -300 & 1050 & -1400 & 630 \\ -300 & 4800 & -18900 & 26880 & -12600 \\ 1050 & -18900 & 79380 & -117600 & 56700 \\ -1400 & 26880 & -117600 & 179200 & -88200 \\ 630 & -12600 & 56700 & -88200 & 44100 \end{bmatrix}$$

Some of the elements of the inverse are orders of magnitude larger than elements of the original matrix. Furthermore, the determinant of the coefficient matrix is nearly zero.

Since each element of the constant vector is the sum of the matrix elements in the corresponding row, the exact solution is:

$$\begin{bmatrix} 1 & 1 & 1 & 1 & 1 \end{bmatrix}$$

Because of the ill conditioning, however, the calculated solution might be something like this:

$$\begin{bmatrix} 1.0001 & 0.99816 & 1.00778 & 0.9884 & 1.0056 \end{bmatrix}$$

Now let us look at a program that uses Hilbert matrices to study the effect of ill conditioning.

BASIC PROGRAM: SOLVING HILBERT MATRICES

The program shown in Figure 4.9 will generate a set of Hilbert matrices and constant vectors corresponding to a solution of:

$$\begin{bmatrix} 1 & 1 & 1 & \dots & 1 \end{bmatrix}$$

Running the Program

Type up the program and execute it. It will run automatically. The program begins with the two equations:

$$\begin{bmatrix} 1 & \frac{1}{2} \\ \frac{1}{2} & \frac{1}{3} \end{bmatrix} \quad \begin{bmatrix} \frac{3}{2} \\ \frac{5}{6} \end{bmatrix}$$

which are solved by utilizing the first version of the Gauss-Jordan method (Figure 4.6). A matrix this small is not ill conditioned, so no problem is apparent. The solution vector is [1 1]. The program then continues with three, four, five, six, and seven equations.

If your BASIC floating-point operations are performed with the usual 32-bit precision, you will start seeing roundoff errors with four equations. Six equations will give accuracy of only two or three significant figures, and the results for seven equations will be meaningless.

Alternately, if you have double-precision, 64-bit floating-point operations, then the situation is very different. In this case, the solution to seventeen simultaneous equations will be given to an accuracy of seven significant figures or better. Thus this program can be used to test the significance of your floating-point package. To run the double-precision version with Microsoft BASIC, add the lines:

```
30 DEFDBL A-H : DEFDBL P-Z
70 M1% = 17
```

Line 30 implicitly defines as double precision all variables starting with the letters A-H and P-Z. Also change each dimension of 8 to 17 in line 80.

```
 10   REM Simultaneous solution of Hilbert matrices
 11   REM by Gaussian elimination, Apr 20, 81
 12   REM identifiers
 14   REM     C1      COEF          solution vector
 15   REM     I2%     INDEX%        work matrix
 16   REM     M1%     MAX%          maximum length
 17   REM     N1%     NROW%         number of rows
 18   REM     N2%     NCOL%         number of columns
 19   REM end of identifiers
 30   REM
 40   A$ = '' ##.###^^^ ''
 50   B$ = ''  = ##.###^^^''
 60   C$ = ''   ##.#####''
 70   M1% = 8
 80   DIM Z(8), A(8,8), C1(8), W(8,1), B(8,8), I2%(8,3)
 90   REM
100   N1% = 2
110   N2% = N1%
120   A(1,1) = 1.0
130   FOR J2% = 2 TO M1%
140      GOSUB 500 : REM input subroutine
150      GOSUB 5000 : REM Gauss—Jordan subroutine
160      REM
170      IF (N1% > 5) THEN 250
180      PRINT ''          Matrix    Constants''
190      FOR I% = 1 TO N1%
200         FOR J% = 1 TO N2%
210            PRINT USING A$; A(I%,J%);
220         NEXT J%
230         PRINT USING B$; Z(I%)
240      NEXT I%
250      PRINT
260      PRINT ''          Solution''
270      PRINT
```

Figure 4.9: Solution of a Set of Ill-Conditioned Equations

```
280      FOR I% = 1 TO N2%
290         PRINT USING C$; C1(I%);
300      NEXT I%
310      PRINT
320      PRINT
330      N1% = N1%+1
340      N2% = N1%
350   NEXT J2%
360   PRINT CHR$(7)
370   GOTO 9999
500   REM
510   REM input the data
520   REM
530   FOR I% = 1 TO N1%
540      A(N1%,I%) = 1.0 / (N1% + I% − 1)
550      A(I%,N1%) = A(N1%,I%)
560   NEXT I%
570   A(N1%,N1%) = 1.0 / (2 ∗ N1% −1)
580   FOR I% = 1 TO N1%
590      Z(I%) = 0
600      FOR J% = 1 TO N1%
610         Z(I%) = Z(I%) + A(I%,J%)
620      NEXT J%
630   NEXT I%
640   RETURN : REM from input routine
5000  REM Gauss−Jordan matrix inversion and solution
      (Continue with lines 5010 to 6130 of Figure 4.6.)
6140  RETURN : REM from Gauss−Jordan subroutine
9999  END
```

Figure 4.9: Solution of a Set of III – Conditioned Equations (cont.)

In the next section we will see another application of the Gauss-Jordan method. We will consider a set of equations in which the number of equations is greater than the number of variables per equation, and we will present a program to compute the "best-fit" solution.

A SIMULTANEOUS BEST FIT

The previous programs in this chapter produce an *exact* fit to a set of linear equations. In each case, the number of unknowns equals the number of independent equations. If, however, there are more unknowns than there are equations, then no unique solution is possible. This situation corresponds to a coefficient matrix having more columns than rows.

There is another case we might consider. Suppose that we want to determine the value of *m* unknowns by an experimental procedure. If *m* independent measurements are obtained, then we can find an exact solution. On the other hand, suppose that the number of independent measurements, *n*, is greater than the number of unknowns, *m*. Then it is possible to calculate a *best-fit* value for the *m* unknowns.

As a two-dimensional analogue, consider the three equations:

$$x + y = 3$$
$$x = 1$$
$$y = 1$$

These three lines do not intersect at the same point. Rather, each pair of lines defines a different point on the *x-y* plane. These points describe a right triangle that has corners at the positions (1, 1), (2, 1), and (1, 2), as shown in Figure 4.10. The best choice for the "intersection" of these

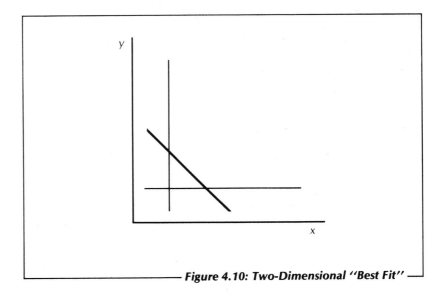

Figure 4.10: Two-Dimensional "Best Fit"

three lines is the location of the centroid of the triangle. Since this point is located at one-third of the distance from the base to the apex, then the best-fit solution to these three lines is:

$$x = 1.3333, \quad y = 1.3333$$

Now that we have illustrated a best-fit solution, we will look at the program that will find the solution.

BASIC PROGRAM: THE BEST-FIT SOLUTION

Make a copy of the program shown in Figure 4.6 and alter it to look like Figure 4.11. Lines 4000-4150 of the matrix multiplication routine given in Figure 3.1 are also needed. This routine is used to convert the rectangular array of coefficients and the constant vector into the square array and vector needed by the Gauss-Jordan routine.

The technique used in this program is essentially the same as that used for *least-squares curve fitting*. Since this topic is covered more fully in the next chapter, it will not be discussed further here.

```
10    REM Simultaneous best—fit solution
11    REM by Gauss—Jordan elimination, Apr 19, 81
12    REM identifiers
14    REM      C1       COEF         solution vector
15    REM      E1%      ERMES%       error flag
16    REM      I2%      INDEX%       work matrix
17    REM      M1%      MAX%         maximum length
18    REM      N1%      NROW%        number of rows
19    REM      N2%      NCOL%        number of columns
20    REM end of identifiers
40    REM
50    A$ = " ##.#### "
60    B$ = " = ##.####"
70    C$ = "  ##.#####"
80    M1% = 8
90    DIM Z(8), A(8,8), C1(8), Y(8), U(8,8)
100   DIM W(8,1), B(8,8), I2%(8,3)
110   REM
```

Figure 4.11: The Best Fit to a Set of Linear Equations

```
120   PRINT "Simultaneous solution by Gauss—Jordan elimination"
130   REM
140   GOSUB 500 : REM get the data
150   GOSUB 4000 : REM square up the matrix
160   GOSUB 5000 : REM get the solution
170   REM
180   IF (N2% > 5) THEN 260
190   PRINT "          Matrix     Constants"
200   FOR I% = 1 TO N1%
210     FOR J% = 1 TO N2%
220       PRINT USING A$; U(I%,J%);
230     NEXT J%
240     PRINT USING B$; Y(I%)
250   NEXT I%
260   PRINT
270   IF (E1% = 1) THEN 340
280   PRINT "          Solution"
290   PRINT
300   FOR I% = 1 TO N2%
310     PRINT USING C$; C1(I%);
320   NEXT I%
330   PRINT
340   GOTO 130
500   REM
510   REM input the data
520   REM
530   PRINT
540   INPUT " How many unknowns"; N2%
550   IF (N2% > M1%) THEN 540
560   IF (N2% < 2) THEN 9999
570   INPUT " How many equations"; N1%
580   IF (N1% < N2%) THEN 570
```

Figure 4.11: The Best Fit to a Set of Linear Equations (cont.)

```
590    FOR I% = 1 TO N1%
600      PRINT "Equation"; I%
610      FOR J% = 1 TO N2%
620        PRINT J%; "  ";
630        INPUT U(I%,J%)
640      NEXT J%
650      INPUT "  C  "; Y(I%)
660    NEXT I%
670    RETURN : REM from input routine
4000   REM U and Y converted to A and Z
4011   REM identifiers
4013   REM     N1%     NROW%     number of rows
4014   REM     N2%     NCOL%     number of columns
4015   REM end of identifiers
4016   REM
4020   FOR K% = 1 TO N2%
4030     FOR L% = 1 TO K%
4040       A(K%,L%) = 0
4050       FOR I% = 1 TO N1%
4060         A(K%,L%) = A(K%,L%) + U(I%,L%) * U(I%,K%)
4070         IF K% <> L% THEN A(L%,K%) = A(K%,L%)
4080       NEXT I%
4090     NEXT L%
4100     Z(K%) = 0
4110     FOR I% = 1 TO N1%
4120       Z(K%) = Z(K%) + Y(I%) * U(I%,K%)
4130     NEXT I%
4140   NEXT K%
4150   RETURN : REM from square
5000   REM Gauss — Jordan matrix inversion and solution
       (Continue with lines 5010 to 6130 of Figure 4.6.)
6140   RETURN : REM from Gauss — Jordan subroutine
9999   END
```

Figure 4.11: The Best Fit to a Set of Linear Equations (cont.)

Running the Best-Fit Program

Type up the program and execute it. First the number of unknowns is requested, then the number of equations. If these have the same value, then a regular, simultaneous solution is performed. However, if there are more equations than unknowns, then the best fit is returned. The three equations that we discussed earlier in this section have only two unknowns. The matrix and constant vector are:

$$\begin{bmatrix} 1 & 1 \\ 1 & 0 \\ 0 & 1 \end{bmatrix} \qquad \begin{bmatrix} 3 \\ 1 \\ 1 \end{bmatrix}$$

Verify that the result is:

$$x = 1.3333, \quad y = 1.3333$$

As another example, consider the electric circuit from the beginning of this chapter. We originally derived three loop-current equations that were solved simultaneously. Suppose, however, that the voltages of the sources were determined experimentally. If the value of the left source was found to be 19 volts, and value of the right source was measured as -5.1 volts, then the three loop equations would be:

$$\begin{bmatrix} 13 & -8 & -3 \\ -8 & 10 & -1 \\ -3 & -1 & 11 \end{bmatrix} \qquad \begin{bmatrix} 19 \\ -5.1 \\ 0 \end{bmatrix}$$

In addition, suppose that the voltage across the horizontal one-ohm resistor was measured to be 1.1 volts. The current flowing through this resistor is loop current 2 minus loop current 3. Consequently, we can now write an additional independent equation, corresponding to an additional row in our matrix. The four equations are:

$$\begin{bmatrix} 13 & -8 & -3 \\ -8 & 10 & -1 \\ -3 & -1 & 11 \\ 0 & 1 & -1 \end{bmatrix} \qquad \begin{bmatrix} 19 \\ -5.1 \\ 0 \\ 1.1 \end{bmatrix}$$

Enter these four equations, with their three unknowns, into the new program. Compare the resulting, best-fit solution:

$$I_1 = 2.8 \, amps, \quad I_2 = 1.8 \, amps, \quad I_3 = 0.94 \, amps$$

to the original solution. This program can be aborted by entering zero for the number of equations.

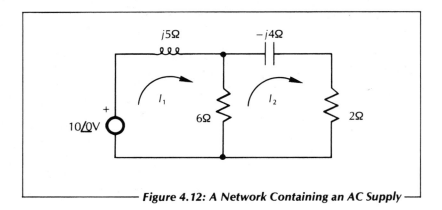

Figure 4.12: A Network Containing an AC Supply

Next we will devise a method—and a BASIC program—for solving simultaneous equations that have complex coefficients (i.e., factors containing imaginary parts). To illustrate this problem we will study a second, more complicated electrical example.

EQUATIONS WITH COMPLEX COEFFICIENTS

Simultaneous equations with complex coefficients occur in the analysis of electrical circuits. Complex numbers are not incorporated into BASIC. Nevertheless, it is rather easy to solve n complex simultaneous equations by converting them into $2n$ equations with real coefficients. The resulting set can then be solved by one of the methods developed previously in this chapter.

Example: An Alternating-Current Electrical Circuit

Consider the electrical circuit shown in Figure 4.12. This circuit is more complicated than Figure 4.1 since it contains an AC power source, an inductor, and a capacitor. The impedance for a resistor is simply the resistance R. However, the impedance for inductors and capacitors is a function of the frequency. The impedance function for the inductor is $j\omega L$, where j is the imaginary operator equal to the square root of -1, ω is the frequency of the AC source in radians per second, and L is the self-inductance in henries. The impedance function for the capacitor is $-j/\omega C$, where C is the capacitance in farads.

For Figure 4.12, the impedance functions are shown next to the corresponding elements. The AC power source is 10 volts (RMS) with a frequency of ω. The phase angle is arbitrarily chosen to be zero. The inductor has an impedance of $j5$ ohms and the capacitor has an impedance of $-j4$ ohms.

We can find the branch currents for this circuit by using the two loop currents and the Kirchhoff voltage law. A clockwise summing around each loop gives:

$$(6 + j5)I_1 - 6I_2 - 10 = 0 \quad \text{(left loop)}$$
$$-6I_1 + (8-j4)I_2 = 0 \quad \text{(right loop)}$$

The corresponding linear equations are:

$$\begin{bmatrix} (6 & + & j5) & (-6 & + & j0) \\ (-6 & + & j0) & (8 & - & j4) \end{bmatrix} \quad \begin{bmatrix} (10 & + & j0) \\ (0 & + & j0) \end{bmatrix}$$

These two equations cannot be directly solved with the programs given previously in this chapter because they contain complex coefficients.

Let us consider a general statement of the two loop equations. The electrical current and the impedance can both be expressed as complex numbers. Consequently, we can write:

$$(AR_{11} + jAI_{11}) (IR_1 + jII_1) + (AR_{21} + jAI_{21}) (IR_2 + jII_2)$$
$$= (VR_1 + jVI_1)$$

$$(AR_{21} + jAI_{21}) (IR_1 + jII_1) + (AR_{22} + jAI_{22}) (IR_2 + jII_2)$$
$$= (VR_2 + jVI_2)$$

where the following symbols are used:

AR_{kl} = real part of coefficient (impedance) k,l
AI_{kl} = imaginary part of coefficient k,l
IR_l = real part of current l
II_l = imaginary part of current l
VR_k = real part of voltage for equation k
VI_k = imaginary part of voltage for equation k

Multiplication of the terms on the left of the above equations produces groups that alternately include the complex operator j.

$$(AR_{11} IR_1 - AI_{11} II_1) + j(AR_{11} II_1 + AI_{11} IR_1) + \ldots$$

If the complex expression on the left is to equal the complex expression on the right, then the real terms on the left must equal the real terms on the right. Similarly, the real coefficients of the imaginary terms on the left must equal the corresponding terms on the right. This approach gives rise to a new set of 2n simultaneous equations that contain only real coefficients.

The first new equation is set equal to the real part of the first constant (voltage) term:

$$AR_{11} - AI_{11} + AR_{12} - AI_{12} = VR_1$$

Notice that the complex conjugates of the original coefficients appear in the new first equation. That is, the original coefficients appear in order, but with alternating signs.

The second new equation is set equal to the imaginary part of the first constant (voltage) term:

$$AI_{11} + AR_{11} + AI_{12} + AR_{12} = VI_1$$

This equation also contains all of the coefficients for the first original equation, but in this case the real and imaginary parts are interchanged. Furthermore, the original signs are utilized.

The complete new equations can be summarized as:

$$\begin{bmatrix} AR_{11} & -AI_{11} & AR_{12} & -AI_{12} \\ AI_{11} & AR_{11} & AI_{12} & AR_{12} \\ AR_{21} & -AI_{21} & AR_{22} & -AI_{22} \\ AI_{21} & AR_{21} & AI_{22} & AR_{22} \end{bmatrix} \begin{bmatrix} VR_1 \\ VI_1 \\ VR_2 \\ VI_2 \end{bmatrix}$$

Substituting the values from Figure 4.12 gives:

$$\begin{bmatrix} 6 & -5 & -6 & 0 \\ 5 & 6 & 0 & -6 \\ -6 & 0 & 8 & 4 \\ 0 & -6 & -4 & 8 \end{bmatrix} \begin{bmatrix} 10 \\ 0 \\ 0 \\ 0 \end{bmatrix}$$

Notice that each original coefficient appears twice in the new matrix. The solution vector for the new set of equations can readily be found by the methods discussed in this chapter. The solution is:

$$\begin{bmatrix} 1.5 & -2.0 & 1.5 & -0.75 \end{bmatrix}$$

which corresponds to the loop currents:

$$I_1 = 1.5 - j2 \quad \text{amps} \quad = \quad 2.5 \ \angle{-53°}$$
$$I_2 = 1.5 - j0.75 \quad \text{amps} \quad = \quad 1.67 \ \angle{-27°}$$

These results can be readily verified by calculating the voltages across each circuit element. For example, if the lower node is chosen to be zero volts, then the voltage of the upper node is equal to the voltage across the 6-ohm resistor:

$$V = 6(I_1 - I_2) = 6(-j1.25) = -j7.5 \text{ volts}$$

Similarly, the voltage across the inductor is:

$$V = j5(I_1) = 10 + j7.5 \text{ volts}$$

A sum of the voltages around the left loop then gives:

$$(-j7.5) + (10 + j7.5) - (10) = 0$$

A similar check can be made on the right loop.

Let us now look at a program that will handle these complex coefficients.

BASIC PROGRAM:
SIMULTANEOUS EQUATIONS WITH COMPLEX COEFFICIENTS

The program given in Figure 4.13 simplifies the solution of simultaneous equations with complex coefficients. Each coefficient of the original n equations is entered only once. Then the program converts the data into a $2n$-by-$2n$ matrix and a constant vector of length $2n$. Up to four complex equations can be solved simultaneously. This number can be increased by changing the value of M1% and the corresponding dimension statements. (The value must be twice the maximum number of equations.) Any of the methods previously developed in this chapter can find the solution. We have selected the Gauss-Jordan technique. Consequently, it might be best to begin with a copy of Figure 4.6.

```
10    REM Simultaneous solution of complex equations
11    REM by Gauss — Jordan elimination, Apr 19, 81
12    REM identifiers
14    REM     A1      ANG      phasor angle
15    REM     C1      COEF     solution vector
16    REM     D4      DR       real array
17    REM     D5      DI       imaginary array
18    REM     E1%     ERMES%   error flag
19    REM     I2%     INDEX%   work matrix
20    REM     I6      IMA      imaginary part
21    REM     M1%     MAX%     maximum length
22    REM     M3      MAG      scalar magnitude
23    REM     N1%     NROW%    number of rows
24    REM     N2%     NCOL%    number of columns
25    REM     R2      RE       real part
26    REM end of identifiers
30    REM
```

Figure 4.13:
Simultaneous Solution of Equations with Complex Coefficients

```
40    A$ = " ##.###^^^^ "
50    B$ = " = ##.###^^^^"
60    C$ = "  ##.#####    ##.#####    ##.#####    ###.#####"
70    M1% = 8
80    DIM Z(8), A(8,8), C1(8), W(8,1), B(8,8), I2%(8,3)
90    DIM D4(4,4), D5(4,4), V(4,2)
100   P8 = 180 / 3.14159
110   REM
120   PRINT "Simultaneous solution with complex coefficients"
130   PRINT "By Gauss—Jordan elimination"
140   GOSUB 500 : REM input routine
150   GOSUB 5000 : REM Gauss—Jordan routine
160   REM
170   IF (N1% > 5) THEN 250
180     PRINT "          Matrix     Constants"
190   FOR I% = 1 TO N1%
200     FOR J% = 1 TO N2%
210       PRINT USING A$; A(I%,J%);
220     NEXT J%
230     PRINT USING B$; Z(I%)
240   NEXT I%
250   PRINT
260   IF (E1% = 1) THEN 400
270   PRINT "   Real      Imaginary    Magnitude    Angle"
280   PRINT
290   FOR I% = 1 TO N2% / 2
300     J% = 2 * I% −1
310     R2 = C1(J%)
320     I6 = C1(J% + 1)
330     M3 = SQR(R2 * R2 + I6 * I6)
340     IF (R2 > 0) THEN A1 = ATN(I6 / R2) * P8
350     IF (R2 = 0) THEN A1 = SGN(I6) * 90
360     IF (R2 < 0) THEN A1 = ATN(I6/R2) * P8 + 180
370     PRINT USING C$; R2, I6, M3, A1
380   NEXT I%
```

Figure 4.13:
Simultaneous Solution of Equations with Complex Coefficients (cont.)

```
390    PRINT
400    GOTO 140 : REM next set of equations
500    REM
510    REM input the data
520    REM
530    PRINT
540    INPUT " How  many  equations"; N1%
550    IF (N1% > M1% / 2) THEN 540
560    IF (N1% < 2) THEN 9999
570    N2% = N1%
580    FOR I% = 1 TO N1%
590      PRINT "Equation"; I%
600      K% = 0
610      L% = 2 * I% - 1
620      FOR J% = 1 TO N1%
630        K% = K% + 1
640        PRINT "Real "; J%; "   ";
650        INPUT D4(I%, J%)
660        A(L%,K%) = D4(I%, J%)
670        A(L% + 1,K% + 1) = D4(I%, J%)
680        K% = K% + 1
690        PRINT "Imag "; J%; "   ";
700        INPUT D5(I%, J%)
710        A(L%,K%) = -D5(I%, J%)
720        A(L% + 1,K% - 1) = D5(I%, J%)
730      NEXT J%
740      INPUT "Real const "; V(I%,1)
750      Z(L%) = V(I%,1)
760      INPUT "Imag const "; V(I%,2)
770      Z(L% + 1) = V(I%,2)
780    NEXT I%
790    PRINT : REM print original matrix
```

Figure 4.13:
Simultaneous Solution of Equations with Complex Coefficients (cont.)

```
800    FOR I% = 1 TO N1%
810      FOR J% = 1 TO N2%
820        PRINT D4(I%,J%); D5(I%,J%);
830      NEXT J%
840      PRINT V(I%,1); V(I%,2)
850    NEXT I%
860    N1% = 2 * N1%
870    N2% = N1%
880    RETURN : REM from input routine
5000   REM Gauss — Jordan matrix inversion and solution
       (Continue with lines 5010 to 6130 of Figure 4.6.)
6140   RETURN : REM from Gauss — Jordan subroutine
9999   END
```

Figure 4.13:
Simultaneous Solution of Equations with Complex Coefficients (cont.)

Running the Program

Type up the program and use it to solve the circuit shown in Figure 4.12. Enter the value of 2 for the number of complex equations. Then enter the coefficients for each equation in turn. Give the real coefficient first, then the imaginary part next. The constant vector is entered in the same way; real part first, then imaginary part. The data for this problem are entered as:

$$6, 5, -6, \quad 0, 10, 0 \text{ (equation 1)}$$
$$-6, 0, \quad 8, -4, \quad 0, 0 \text{ (equation 2)}$$

The original input data are printed out, and the new 2n-by-2n matrix is given. Then, the solution is given in both rectangular and polar forms.

The polar magnitude is calculated by finding the square root of the sum of the squares of the rectangular components. The phasor angle is determined from the arctangent. This function returns an angle in the first quadrant, (0 to 90 degrees) if the argument is positive. If the argument is negative, the result is in the fourth quadrant (0 to −90 degrees). But if the x coordinate of the angle is negative, then 180 degrees must be added to the angle to place the result into the correct quadrant.

We have now seen several methods and variations that are adequate for solving small linear matrices (i.e., solving small sets of simultaneous equations). The last topic of this chapter will be an iterative method for solving large matrices or nonlinear matrices.

THE GAUSS-SEIDEL ITERATIVE METHOD

The Gauss elimination and Gauss-Jordan methods we considered previously are not suitable for solving very large matrices. More and more multiplication and subtraction operations are performed as the number of equations increases. The resulting roundoff error can produce a meaningless solution.

The Gauss-Seidel method finds the solution to a set of equations by an iterative technique. An initial approximation is repeatedly refined until the result is acceptably close to the solution. Since each approximation depends only on the previous approximation, roundoff error does not accumulate. An added feature is that the equations do not have to be linear.

Consider the three loop-current equations derived from Figure 4.1:

$$13I_1 - 8I_2 - 3I_3 = 20$$
$$-8I_1 + 10I_2 - I_3 = -5$$
$$-3I_1 - I_2 + 11I_3 = 0$$

We can solve the first equation for I_1:

$$I_1 = \frac{20 + 8I_2 + 3I_3}{13}$$

in terms of the other two unknowns. Then if we choose first approximations of zero for I_2 and I_3, we obtain a value of 1.54 for I_1. The second equation is then solved for the second variable:

$$I_2 = \frac{8I_1 + I_3 - 5}{10}$$

Substituting the current values of 1.54 for I_1 and zero for I_3 produces a value of 0.73 for I_2. The third equation is similarly solved for the third variable:

$$I_3 = \frac{3I_1 + I_2}{11}$$

The current values of 1.54 for I_1 and 0.73 for I_2 give a value of 0.486 for I_3. The process is now repeated. The values of 0.73 for I_2 and 0.486 for I_3 are used to obtain a better value for I_1. After about 20 complete iterations, the values are correct to three significant figures. The following table gives the initial values in the sequence:

I_1	I_2	I_3
0	0	0
1.54	0.73	0.486
2.1	1.23	0.685
2.45	1.53	0.808
2.67	1.71	0.883

There are several potential problems with the Gauss-Seidel method. First of all, the process might not converge. That is, successive values may drift further and further from the correct solution. Consider, for example, the previous three equations. If they are written in reverse order, then we will derive the expressions:

$$I_1 = (11 I_3 - I_2) / 3$$
$$I_2 = (13 I_1 - 3 I_3 - 20) / 13$$
$$I_3 = 10 I_2 - 8 I_1$$

First approximations of zero are chosen, as before. However, the subsequent values are clearly diverging:

I_1	I_2	I_3
0	0	0
0	-1.5	-15
-57	-54	-92
-318	-299	-440

The problem in this second case is that the largest values are not located on the major diagonal. The solution is to interchange rows to bring the largest element into the pivot position.

Finally, let us investigate a BASIC implementation of the Gauss-Seidel method. We will make note of a couple of interesting features in this program:

- the differences between *relative* and *absolute criteria* in **IF .. THEN** decisions
- the meaning and use of *point relaxation*.

BASIC PROGRAM: THE GAUSS-SEIDEL METHOD

Surprisingly, the choice of a first approximation is not too important. An additional matter to be considered, however, is the criterion for convergence. The program shown in Figure 4.14 can be used to explore the Gauss-Seidel method for the solution of linear simultaneous equations. Most of the program can be copied from the Gauss elimination method given in Figure 4.4.

The pivot-interchange routine used in the Gauss elimination program is incorporated here for the same purpose. Rows are interchanged to place the largest element on the major diagonal.

An absolute, rather than a relative, criterion is used to determine convergence:

IF (ABS(C2 — C1(J%)) > T1) **THEN** …

In this expression, C2 is the new value of the coefficient, C1 is the original value, and T1 is the desired tolerance or accuracy. Normally, it is better to choose a relative criterion:

IF (ABS(1 — C1(J%) / C2) > T1) **THEN** …

In this case, however, we must insure that the variable C2 will never be zero. One way to do this would be to use the form:

IF (ABS(C2 — C1(J%)) > ABS(T1 * C2)) **THEN** …

Sometimes the successive approximations jump about wildly. One feature of the program given in Figure 4.14 will reduce this tendency. If two successive approximations differ in sign, then the step size is cut in half.

Another feature of the Gauss-Seidel program shown in Figure 4.14 is known as *point relaxation*. With this technique, each selected value is a function of the previous iteration, the calculated value, and a relaxation factor, lambda. If C1(J%) is the previous value and C2 is the calculated value, then the actual next value becomes:

C1(J%) = L2 * C2 + (1.0 — L2) * C1(J%)

The value of lambda, L2, can range from 0 to 2.

```
10    REM Solution by Gauss—Seidel method, Apr 19, 81
11    REM
12    REM identifiers
14    REM     C1      COEF        solution vector
15    REM     E1%     ERMES%      error flag
16    REM     L       LAMBDA      relaxation factor
17    REM     M1%     MAX%        maximum length
```

Figure 4.14: Solution of Linear Equations by the Gauss-Seidel Method

```
18    REM      N1%     NROW%      number of rows
19    REM      N2%     NCOL%      number of columns
20    REM end of identifiers
30    REM
40    A$ = " ##.###^^^^ "
50    B$ = "  = ##.###^^^^"
60    C$ = "  ##.#####"
70    M1% = 8
80    DIM Z(8), A(8,8), C1(8), W(8), B(8,8)
90    PRINT "Simultaneous solution by Gauss—Seidel"
100   GOSUB 500 : REM input subroutine
110   GOSUB 5000 : REM Gauss elimination
120   REM
130   IF (N1% > 5) THEN 210
140   PRINT "           Matrix     Constants"
150   FOR I% = 1 TO N1%
160     FOR J% = 1 TO N2%
170       PRINT USING A$; A(I%,J%);
180     NEXT J%
190     PRINT USING B$; Z(I%)
200   NEXT I%
210   PRINT
220   IF (E1% = 1) THEN 290
230   PRINT "        Solution"
240   PRINT
250   FOR I% = 1 TO N2%
260     PRINT USING C$; C1(I%);
270   NEXT I%
280   PRINT
290   GOTO 100 : REM next set of equations
500     REM
510     REM input the data
520     REM
530   PRINT
```

Figure 4.14:
Solution of Linear Equations by the Gauss-Seidel Method (cont.)

```
540    INPUT " How many equations"; N1%
550    IF (N1% > M1%) THEN 540
560    IF (N1% = 0) THEN 9999
570    N2% = ABS(N1%)
580    IF (N1% > 0) THEN 610
590        N1% = -N1% : REM use prior matrix
600        GOTO 690
610    FOR I% = 1 TO N1%
620      PRINT "Equation"; I%
630      FOR J% = 1 TO N2%
640          PRINT J%; "   ";
650          INPUT A(I%, J%)
660      NEXT J%
670      INPUT "  C "; Z(I%)
680    NEXT I%
690    INPUT "Relaxation factor"; L2
700    IF (L2 < 0 OR L2 > 2) THEN 690
710    RETURN : REM from input subroutine
5000   REM Simultaneous solution by Gauss—Seidel method
5010   REM Apr 14, 81
5011   REM
5012   REM identifiers
5014   REM      B1      BIG
5015   REM      C1      COEF      solution vector
5016   REM      E1%     ERMES%    error flag
5017   REM      H1      HOLD
5018   REM      J3%     OVER%
5019   REM      L2      LAMBDA    relaxation factor
5020   REM      N2%     NCOL%     number of columns
5021   REM      S6      SUM
5022   REM      T1      TOL       tolerance
5023   REM end of identifiers
5030   REM
5080   E1% = 0
```

Figure 4.14:
Solution of Linear Equations by the Gauss-Seidel Method (cont.)

```
5090    T1 = .0001
5100    FOR I% = 1 TO N2%
5110        FOR J% = 1 TO N2%
5120            B(I%,J%) = A(I%,J%)
5130        NEXT J%
5140    NEXT I%
5150    FOR I% = 1 TO N2% − 1
5160        B1 = ABS(B(I%,I%))
5170        L% = I%
5180        FOR J% = I% + 1 TO N2%
5190            IF (ABS(B(J%,I%)) < B1) THEN 5220
5200            B1 = ABS (B(J%,I%))
5210            L% = J%
5220        NEXT J%
5230        IF (B1 = 0) THEN 5540
5240        IF (L% = I%) THEN 5330
5250        FOR J% = 1 TO N2%
5260            H1 = B(L%,J%)
5270            B(L%,J%) = B(I%,J%)
5280            B(I%,J%) = H1
5290        NEXT J%
5300        H1 = W(L%)
5310        W(L%) = W(I%)
5320        W(I%) = H1
5330    NEXT I%
5340    IF (B(N2%,N2%) = 0) THEN 5540
5350    FOR I% = 1 TO N2%
5360        C1(I%) = 0 : REM initial value
5370    NEXT I%
5380    I% = 0
5390    I% = I% + 1
5400        J3% = 1
5410        FOR J% = 1 TO N2%
5420            S6 = Z(J%)
```

Figure 4.14:
Solution of Linear Equations by the Gauss-Seidel Method (cont.)

```
5430        FOR K% = 1 TO N2%
5440          IF (J% <> K%) THEN S6 = S6 — B(J%,K%) * C1(K%)
5450        NEXT K%
5460        C2 = S6 / B(J%,J%)
5470        IF (ABS(C2 — C1(J%)) < T1) THEN 5560
5480          J3% = 0
5490          IF (C2 * C1(J%) < 0) THEN C2=(C1(J%)+C2)/2
5500        C1(J%) = L2 * C2 + (1 — L2) * C1(J%)
5510        PRINT I%; ''; COEF(''; J%; '') =''; C1(J%)
5520      NEXT J%
5530    IF (I% < = 100 AND J3% = 0) THEN 5390
5540    E1% = 1
5550    PRINT ''No solution''
5560    RETURN : REM from Gauss—Seidel subroutine
9999    END
```

Figure 4.14: Solution of Linear Equations by the Gauss-Seidel Method (cont.)

Running the Gauss-Seidel Program

The **PRINT** statement at line 5510 displays the intermediate values as they are calculated. You might want to remove this line when you understand how the program works.

Generate a copy of Figure 4.14 and execute it. You will be asked for the number of equations. Enter the value of 3. Then enter the three equations from the electric circuit shown in Figure 4.1. The order of the equations is now immaterial since we have incorporated a row-interchange routine. You will next be asked for the relaxation factor. Give a value of 1.0. Each successive iteration will be printed out following the iteration number.

Convergence will occur after about 18 iterations. The program will again ask for the number of equations. Give a value of −3 this time. The minus sign indicates that the equations from the previous step are to be reused.

You can repeatedly run the program, trying different values for the relaxation factor. The following table shows the dependence of number of iterations on the choice of the relaxation factor.

Lambda	Iterations
0.8	31
1.0	20
1.2	11
1.3	9
1.4	12
1.5	15
1.8	44

Next, enter the 2-by-2 Hilbert matrix:

$$\begin{bmatrix} 1.0 & 0.5 \\ 0.5 & 0.3333 \end{bmatrix} \quad \begin{bmatrix} 1.5 \\ 0.83333 \end{bmatrix}$$

You will find that the optimum relaxation factor occurs at a value of 1.4, about the same as for the previous set of equations. Yet, for others sets of equations, the optimum value of lambda might be 1.0 or 0.8.

Clearly, the Gauss-Seidel technique is not as automatic as the others we have considered. But it should be considered for solving large numbers of linear equations, or for solving sets of nonlinear equations.

SUMMARY

We have studied a number of methods for solving simultaneous equations, each method suited to a different situation. We have presented BASIC programs to carry out the algorithms of each of these methods. We have also investigated several special cases: multiple constant vectors; ill-conditioned equations; best-fit solutions for an "overdetermined" equation system; and equations with complex coefficients. In the programs for these special cases we have seen an abundance of new and powerful features of BASIC programming.

EXERCISES

4-1: *Show that the solution to the following set of linear equations:*

$$4x_1 + 3x_2 + 2x_3 + x_4 = 10$$
$$-1x_1 + 4x_2 - 2x_3 - x_4 = 0$$
$$-2x_2 + 4x_3 + 3x_4 = 5$$
$$-2x_1 - x_2 + x_3 + 4x_4 = 2$$

is [1 1 1 1].

4-2: *Show that the solution to the equations:*

$$5x_1 + 2x_2 + x_3 + 3x_4 = 6$$
$$4x_1 + 3x_2 + x_3 - 2x_4 = -4$$
$$3x_1 + x_2 + 2x_3 + x_4 = 2$$
$$-x_1 + x_2 + 5x_3 + 3x_4 = -1$$

is [2 −3 −1 1].

4-3: *Interchange the inductor and the capacitor in Figure 4.12 and find the new loop currents.*

$$I_1 = 2.39 + j1.71 = 2.9/35.5$$
$$I_2 = 1.83 + j0.117 = 1.187/3.58$$

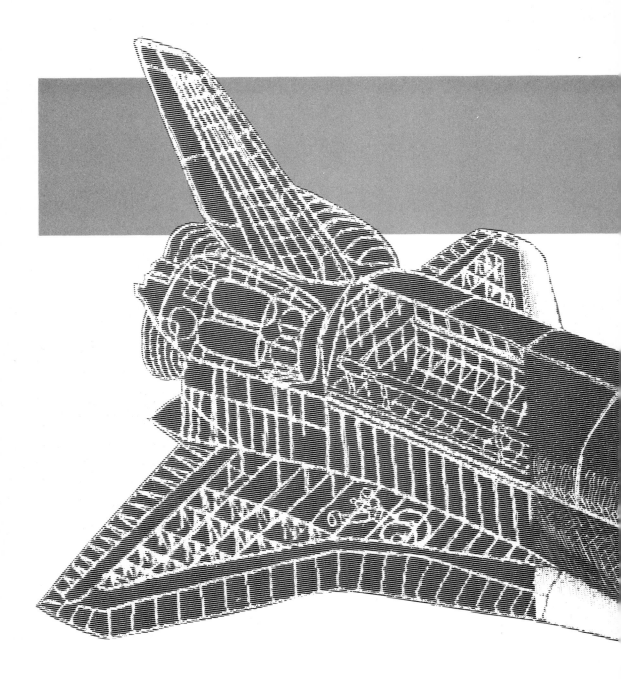

CHAPTER 5

Development of a Curve-Fitting Program

INTRODUCTION

In this chapter we will develop a least-squares curve-fitting program. In particular, we will develop a computer program for finding the best straight line that can represent a set of x-y data. This program will generate the data, calculate the desired equation, print out the results, produce a plot of the data, and supply a measure of the correlation between x and y. A sorting routine will be added in the following chapter to allow handling of real experimental data. Although the resulting program will be large and complex, we will not program all of the parts at one time. Rather, we will use a modular, top-down approach. Only a small portion of the program will be written at first. This part will be checked by actually running it. Another portion of the code will then be added, and it, too, will be checked by running the new program. In this way, a relatively large program can be developed in a logical fashion. Each step will be tested along the way. If an error appears, it will most likely be found in the most recently added portion.

As we develop the different parts of this program, we will be discussing a number of algorithms and their implementations. These include:

- the use of a random function and a "fudge factor" to simulate scattered-line experimental data

- a subroutine for plotting graphs on a regular character-oriented terminal or on a line printer

- a least-squares curve-fitting routine, using differential calculus to arrive at the slope and y-intercept of the actual fitted data

- a simple and elegant method for integrating the correlation coefficient into our program.

THE MAIN PROGRAM

The first thing we will do is write the main program with the input and output routines. The main program will always contain as little as possible: the dimension statement and the calls to the various subroutines. We will consider two versions of the main program—one that uses a built-in random function, and another that simulates that function.

First Version: Using the Built-in RND Function

Create the BASIC source program shown in Figure 5.1. Use a descriptive file name such as:

> CFIT1

for the first version. The program begins by defining the format of the output (A$) and the maximum length of the vectors (M1%). The arrays are dimensioned next. Two subroutines are called. One starting at line 500 provides the input, and the other, at line 1000, performs the output.

```
10    REM Linear least—squares fit, Apr 19, 1981
11    REM
12    REM identifiers
14    REM     F2      FUDGE       fudge factor
15    REM     M1%     MAX%        maximum length
16    REM     N1%     NROW%       number of rows
17    REM end of identifiers
30    REM
40    A$ = " ###   ##.#   ###.##   ###.##"
```

Figure 5.1: The Beginning of a Curve-Fitting Program

```
  50   REM
  60   M1% = 35
  70   DIM X(35), Y(35)
  80   REM
  90   PRINT
 100   REM
 110   GOSUB 500 : REM get the data
 120   REM
 130   REM
 140   REM
 150   GOSUB 1000 : REM Print results
 160   REM
 170   GOTO 110 : REM next fit
 500   REM get the data
 510   A = 2
 520   B = 5
 530   INPUT "Fudge"; F2
 540   IF (F2 < 0) THEN 9999
 550   INPUT "How many points"; N1%
 560   FOR I% = 1 TO N1%
 570      J% = N1% + 1 - I%
 580      X(I%) = J%
 590      Y(I%) = (A + B * J%) * (1 + (2 * RND(1) - 1) * F2)
 600   NEXT I%
 610   REM
 620   RETURN : REM from input routine
1000   REM print results
1010   PRINT "         X         Y"
1020   FOR I% = 1 TO N1%
1030      PRINT USING A$; I%; X(I%), Y(I%)
1040   NEXT I%
1050   RETURN
9999   END
```

Figure 5.1: The Beginning of a Curve-Fitting Program (cont.)

The random number generator is called at line 590 with an argument of unity. It may be necessary to change this argument to zero for your particular dialect of BASIC.

The input routine will initially generate a set of x-y points using a random number generator. In a later chapter we will alter this routine to read data embedded in the program. Another possibility is to read the data from a disk file.

Second Version: Simulating Function RND

The input subroutine at line 500 calls a function named RND to generate a randomly scattered straight line. This function is usually provided with BASIC. However, if your BASIC does not include function RND, you will want to use the program shown in Figure 5.2, which contains the random number generator shown in Figure 2.5, rather than the program shown in Figure 5.1.

```
10    REM Linear least—squares fit, Apr 19, 1981
11    REM
12    REM identifiers
14    REM     F2      FUDGE      fudge factor
15    REM     M1%     MAX%       maximum length
16    REM     N1%     NROW%      number of rows
17    REM     S2      SEED       for random numbers
18    REM end of identifiers
30    REM
40    A$ = '' ###   ##.#   ###.##   ###.##''
50    S2 = 4
60    M1% = 35
70    DIM X(35), Y(35)
80    REM
90    PRINT
100   REM
110   GOSUB 500 : REM get the data
120   REM
130   REM
140   REM
150   GOSUB 1000 : REM print results
```

Figure 5.2: Alternate Routines Including a Random Number Generator

```
160   REM
170   GOTO 110 : REM next fit
500   REM get the data
510   A = 2
520   B = 5
530   INPUT "Fudge"; F2
540   IF (F2 < 0) THEN 9999
550   INPUT "How many points"; N1%
560   FOR I% = 1 TO N1%
570      J% = N1% + 1 − I%
580      X(I%) = J%
585      GOSUB 9800 : REM random number
590      Y(I%) = (A + B * J%) * (1 + (2 * R1 − 1) * F2)
600   NEXT I%
610   REM
620   RETURN : REM from input routine
1000  REM print results
1010  PRINT "          X          Y"
1020  FOR I% = 1 TO N1%
1030     PRINT USING A$; I%; X(I%), Y(I%)
1040  NEXT I%
1050  RETURN
9800  REM
9801  REM identifiers
9803  REM      R1      RAN          random number
9804  REM      S2      SEED         random number seed
9805  REM end of identifiers
9806  REM
9810  R1 = (S2 + 3.14159)^5.04
9820  R1 = R1 − INT(R1)
9830  S2 = R1
9840  RETURN : REM from random number generator
9999  END
```

Figure 5.2:
Alternate Routines Including a Random Number Generator (cont.)

Let us now look more closely at the input subroutine and the algo-rithm for simulating experimental data.

The Scattering Algorithm

The input subroutine starting at line 500 generates a straight line with an intercept *(A)* of 2 and a slope *(B)* of 5. That is, it generates a set of data in the arrays *x* and *y*, corresponding to the line:

$$y = 2 + 5x$$

A random number generator is then used to move the points off the line according to the variable F2 (the "fudge factor"). If the value of F2 is zero, a perfectly straight line is generated. If F2 has a value of 0.2, on the other hand, the points will be displaced to a maximum of 20 percent from the line.

It would be more realistic to generate Gaussian random numbers than uniformly distributed random numbers in this application. However, our chosen method is much faster. Furthermore, we will be removing this part of the program when real data are incorporated into the curve-fitting routine.

The scattering algorithm works in the following way. Function RND returns a real number in the range of 0 to 1. This value is doubled to give a range of 0 to 2. The subtraction of 1 sets the range from minus 1 to plus 1. Finally, this result is multiplied by the fudge factor, F2, and added to unity to give the desired range.

Running the Main Program

Compile the program and try it out. You will be asked to input a value for the fudge factor. Give a value of zero the first time. Then you will be asked for the number of points. Give the value of 9. The result will be three columns of numbers, as shown in Figure 5.3. Notice that the elements of the array X are generated in descending order. We will be able

```
          Fudge?  0
          How many points?  9
                X        Y
          1    9.0     47.00
          2    8.0     42.00
          3    7.0     37.00
          4    6.0     32.00
          5    5.0     27.00
          6    4.0     22.00
          7    3.0     17.00
          8    2.0     12.00
          9    1.0      7.00
```

Figure 5.3: First Run of the Curve-Fitting Program with the Fudge Factor Zero

to reorder the arrays X and Y when we add a sorting routine in the next chapter.

At the completion of this task, the program asks for another value for the fudge factor. This time, respond with 0.2. The resulting x values should be the same as the previous values. The y values, however, will be somewhat larger or smaller than their previous values. When this step is finished, you will be asked for another value for F2. Enter a negative number this time to terminate the program.

Now that you have a working program, you can begin to add new features. Always keep copies of the previous versions. Then if you have trouble with a new version, you can return to the previous version and start again.

A PRINTER PLOTTER ROUTINE

Next we will add a routine for plotting the results on an ordinary computer terminal. Large computers are typically provided with a digital plotter and perhaps a graphic video terminal. With these devices, it is possible to display data to a high degree of precision. Unfortunately, this approach is usually time-consuming. Furthermore, small computers may not have such devices.

As an alternative, experimental data can be displayed on a regular character-oriented terminal or on a line printer by using plus (+) and star (*) symbols. The resulting plot will be a crude representation of the actual data. However, this plot may be useful for finding gross errors in programming as well as incorrectly entered data. Furthermore, the resulting plot can be viewed immediately along with the computational results, whereas there may be a delay in obtaining plotter output from a large computer.

The BASIC program given in Figure 5.1 is shown again in Figure 5.4 with a plotting subroutine added. This will allow us to plot two dependent variables (Y and Y2) as a function of a third independent variable (X). This subroutine plots the independent variable vertically rather than in the usual horizontal direction. The dependent variables are then displayed horizontally. That is, the graph is rotated clockwise a quarter turn from the usual axis orientation. We will use this plot to display uniformly spaced simple functions. However, multivalued functions can also be displayed. Furthermore, the values of the independent variable need not be uniformly spaced.

If your screen is narrower than 64 characters wide, you will have to reduce the width of the plot accordingly. For example, change the value of L6% to 31 at line 7040 and reduce the size of B$, B1$, and P3$.

```
10    REM Linear least—squares fit, Apr 19, 1981
11    REM
12    REM identifiers
14    REM     F2        FUDGE        fudge factor
15    REM     L3%       NLIN%        number of plot lines
16    REM     M1%       MAX%         maximum length
17    REM     N1%       NROW%        number of rows
18    REM     Y2        YCALC        calculated Y
19    REM end of identifiers
30    CLEAR 200 : REM set string space
40    A$ = " ###   ##.#   ###.##   ###.##"
50    REM
60    M1% = 35
70    DIM X(35), Y(35), Y2(35)
80    REM
90    PRINT
100   REM
110   GOSUB 500 : REM get the data
120   REM
130   REM
140   GOSUB 5000 : REM simulate a fit
150   GOSUB 1000 : REM print results
160   GOSUB 7000 : REM plot the data
170   GOTO  110 : REM next fit
500   REM get the data
510   A = 2
520   B = 5
530   INPUT "Fudge"; F2
540   IF (F2 < 0) THEN 9999
550   INPUT "How  many  points"; N1%
560   FOR I% = 1 TO N1%
570      J% = N1% + 1 — I%
580      X(I%) = J%
590      Y(I%) = (A + B * J%) * (1 + (2 * RND(1) — 1) * F2)
600   NEXT I%
```

Figure 5.4: A Plotting Routine is Added

```
 610   L3% = (N1% − 1) * 2 + 1
 620   RETURN : REM from input routine
1000   REM print results
1010   PRINT "         X      Y      Y Calc"
1020   FOR I% = 1 TO N1%
1030      PRINT USING A$; I%; X(I%), Y(I%), Y2(I%)
1040   NEXT I%
1050   RETURN : REM from printout
5000   REM Simulate curve fit with straight line
5010   REM
5020   FOR I% = 1 TO N1%
5030      Y2(I%) = A + B * X(I%)
5040   NEXT I%
5050   RETURN : REM from simulated fit
7000   REM Plot Y and Y2 as a function of X, Apr 20, 81
7010   REM
7011   REM identifiers
7013   REM    L3%     NLIN%        number of plot lines
7014   REM    L4%     LBUF%        line buffer
7015   REM    L6%     LINEL%       line length
7016   REM    N1%     NROW%        number of rows
7017   REM    N4%     NTRY%        entry flag
7019   REM    X2      XLABEL       x label
7020   REM    X3      XHIGH
7021   REM    X4      XLOW
7022   REM    X5      XNXT         next x line
7023   REM    X6      XSCALE
7024   REM    Y2      YCALC        calculated Y
7025   REM    Y3      YHIGH
7026   REM    Y4      YLOW
7027   REM    Y5      YSCALE
7028   REM end of identifiers
7029   REM arrays L4% and V2 are dimensioned at this point
7030   L6% = 51 : B$ = "^            " : REM plot scale
7035   B1$ = "######.#   " : P3$ = "       "
```

Figure 5.4: A Plotting Routine is Added (cont.)

```
7040    PRINT : PRINT : PRINT
7050    IF (L3% < 3) THEN L3% = 15 : REM default plot length
7060    IF (N4%) THEN 7100
7070    N4% = 1
7080    DIM L4%(51), V2(6)
7090    DEF FNK4(P) = (P − Y4) / Y5 + 1
7100    X4 = X(1)
7110    X3 = X(N1%)
7120    Y4 = Y(1)
7130    Y3 = Y4
7140    X6 = (X3 − X4)/(L3% − 1)
7150    IF (X6 > 0) THEN W2 = 1
7160    IF (X6 < = 0) THEN W2 = −1
7170    FOR I% = 1 TO N1%
7180        IF (Y(I%) < Y4) THEN Y4 = Y(I%)
7190        IF (Y2(I%) < Y4) THEN Y4 = Y2(I%)
7200        IF (Y(I%) > Y3) THEN Y3 = Y(I%)
7210        IF (Y2(I%) > Y3) THEN Y3 = Y2(I%)
7220    NEXT I%
7230    Y5 = (Y3 − Y4)/(L6% − 1)
7240    FOR I% = 1 TO L6%
7250        L4%(I%) = 0
7260    NEXT I%
7270    X2 = X4
7280    J4% = 0
7290    L% = 1
7300    K2% = FNK4(Y(1))
7310    L4%(K2%) = 1
7320    K2% = FNK4(Y2(1))
7330    L4%(K2%) = 2
7340    FOR I% = 2 TO L3%
7350        X5 = X4 + X6*(I%−1)
7360        IF (J4% = 1) THEN 7460
7370        L% = L% + 1
7380        IF ((X(L%) − (X5 − X6*.5))*W2< =0) THEN 7410
```

Figure 5.4: A Plotting Routine is Added (cont.)

```
7390       GOSUB 7820
7400       GOTO 7470
7410       K2% = FNK4(Y(L%))
7420       L4%(K2%) = 1
7430       K2% = FNK4(Y2(L%))
7440       L4%(K2%) = 2
7450       GOTO 7370
7460       PRINT " —"
7470       IF ((X(L%) − (X5 + X6*.5))*W2 < = 0) THEN 7500
7480       J4% = 1
7490       GOTO 7560
7500       J4% = 0
7510       K2% = FNK4(Y(L%))
7520       L4%(K2%) = 1
7530       K2% = FNK4(Y2(L%))
7540       L4%(K2%) = 2
7550       X2 = X5
7560     NEXT I%
7570     IF (L% > = N1%) THEN 7640
7580     L% = L% + 1
7590     K2% = FNK4(Y(L%))
7600     L4%(K2%) = 1
7610     K2% = FNK4(Y2(L%))
7620     L4%(K2%) = 2
7630     GOTO 7570
7640     GOSUB 7820
7650     Y6 = Y5 * 10
7660     V2(1) = Y4
7670     FOR J% = 1 TO 4
7680        V2(J% + 1) = V2(J%) + Y6
7690     NEXT J%
7700     V2(6) = Y3
7710     PRINT P3$; "    ";
7720     FOR J% = 1 TO 6
7730        PRINT B$;
```

Figure 5.4: A Plotting Routine is Added (cont.)

```
7740    NEXT J%
7750    PRINT
7760    PRINT "  ";
7770    FOR J% = 1 TO 6
7780        PRINT USING B1$; V2(J%);
7790    NEXT J%
7800    PRINT
7810    RETURN
7820    FOR J% = 51 TO 1 STEP −1
7830        IF (L4%(J%)) THEN 7870
7840    NEXT J%
7850    PRINT
7860    RETURN
7870    J5% = J%
7880    L$ = ""
7890    FOR J% = 1 TO J5%
7900        IF (L4%(J%) = 0) THEN L$ = L$ + " "
7910        IF (L4%(J%) = 1) THEN L$ = L$ + "+"
7920        IF (L4%(J%) = 2) THEN L$ = L$ + "*"
7930    NEXT J%
7940    PRINT X2; P3$; L$
7950    FOR J% = 1 TO L6%
7960        L4%(J%) = 0
7970    NEXT J%
7980    L$ = ""
7990    RETURN : REM end of plot
9999    END
```

Figure 5.4: A Plotting Routine is Added (cont.)

Running the Plotter Routine

Make a copy of the previous version of your source program and give it a distinctive name such as CFIT2. Add the plotting routine starting at line 7000. In addition, you must insert a call to the plotting routine at line 160. Furthermore, the length of the plot is set with the statement at line 610. These two lines were previously set aside for this purpose through the use of **REM** statements. Some BASICs require a **CLEAR** statement to set string space. This statement appears on line 30. Other BASICs ignore this command. Still others forbid its use.

The plotting subroutine is programmed to plot two dependent variables, Y and Y2. Therefore, it is necessary to add another subroutine starting at line 5000. This routine will generate the vector Y2 that matches the original straight line with a slope of 5 and an intercept of 2. This temporary routine will be replaced by a much more significant subroutine later in the chapter, after we have studied the least-squares curve-fitting algorithm. The call to this subroutine is placed at line 140.

Run this version to try it out. Again, you will be asked to give a value for the fudge factor. Respond with an answer of 0.2. This will produce four columns of data, including the values for Y2. The printer plot that follows the tabular data will show two different sets of data. Star symbols are used to represent the values of Y2. They will form a nearly perfect straight line. In addition, '+' symbols represent the values of Y. They will be scattered on both sides of the calculated values (Y2) as shown in Figure 5.5. If the two symbols are coincident for a given value of X, only the star is given.

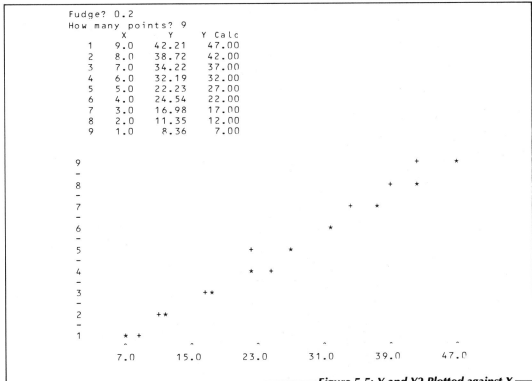

Figure 5.5: Y and Y2 Plotted against X

When the program asks for another value for the fudge factor, respond with a zero. In this case, both Y and Y2 will have the same values. Since the two lines lie one on top of the other, only one line will be shown. Finally, give a negative value for the fudge factor to terminate the program.

Now that we have written and tested both the main program and a procedure for plotting curves, we are ready to move on to the real topic of this chapter.

THE CURVE-FITTING ALGORITHM

Though it may appear from Figure 5.5 that we have completed our curve-fitting program, we have actually not done so. We have simply generated a set of data (Y2) that corresponds to our original line. The time has come to derive the algorithm for a linear, least-squares procedure. We will first introduce a new vector, **r**, which contains the *residuals*.

For each experimental point, corresponding to an x-y pair, there will be an element of **r** that represents the difference between the corresponding calculated value $\hat{y}$ (pronounced y-hat) and the original value of y. This can be expressed mathematically as:

$$r_i = \hat{y}_i - y_i \tag{1}$$

Occasionally, a point will coincide with the calculated curve. But in general, about half of the x-y points will lie on one side of the fitted curve, resulting in a positive value for r. The remaining points will lie on the other side of the curve and give negative values for r. The sum of these residuals should be close to zero.

The least-squares, curve-fitting criterion is that the sum of the squares of the residuals be minimized. The square of each residual will be positive; therefore, the sum of the residuals squared (SRS) will be a positive number. This criterion can be expressed as:

$$SRS = \sum_{i-1}^{n} r_i^2 = \text{minimum} \tag{2}$$

where n is the number of x-y points (and the length of the vectors **x**, **y**, and $\hat{\mathbf{y}}$).

By combining Equation 1 with the curve-fitting equation:

$$\hat{y}_i = A + Bx_i \tag{3}$$

we get

$$r_i = A + Bx_i - y_i \tag{4}$$

and

$$SRS = \sum_{i=1}^{n} r_i^2 = \sum_{i=1}^{n} (A + Bx_i - y_i)^2 \tag{5}$$

The problem is reduced to finding the values of A and B so that the summation of Equation 5 is minimized. We do this with differential calculus. We take the derivative of Equation 5 with respect to each variable (A and B in this case) and set the result to zero.

$$\frac{\delta \Sigma r_i^2}{\delta A} = 0 \quad \text{and} \quad \frac{\delta \Sigma r_i^2}{\delta B} = 0 \tag{6}$$

Substitution of Equation 5 into Equations 6 gives:

$$\frac{\delta \Sigma (A + Bx_i - y_i)^2}{\delta A} = 0 \tag{7}$$

and

$$\frac{\delta \Sigma (A + Bx_i - y_i)^2}{\delta B} = 0 \tag{8}$$

which is equivalent to:

$$\frac{2 \Sigma (A + Bx_i - y_i) \, \delta \Sigma (A + Bx_i - y_i)}{\delta A} = 0 \tag{9}$$

and

$$\frac{2 \Sigma (A + Bx_i - y_i) \, \delta \Sigma (A + Bx_i - y_i)}{\delta B} = 0 \tag{10}$$

Since B, x, and y are not functions of A, and the derivative of A with respect to itself is unity, Equation 9 reduces to:

$$\Sigma A + \Sigma Bx_i = \Sigma y_i \tag{11}$$

Similarly, A, x and y are not functions of B. Therefore, Equation 10 becomes:

$$\Sigma Ax_i + \Sigma Bx_i^2 = \Sigma x_i y_i \tag{12}$$

A and B are constants. Therefore they can be factored from the summation step. Equations 7 and 8 can then be expressed as:

$$An + B\Sigma x_i = \Sigma y_i \tag{13}$$

and

$$A\Sigma x_i + B\Sigma x_i^2 = \Sigma x_i y_i \tag{14}$$

We thus have reduced the problem of finding a straight line through a set of x-y data points to one of solving two simultaneous equations (13 and 14). Both these equations are linear in the unknowns A and B. (x, y, and n are, of course, the original data.) The simultaneous solution can be obtained by using Cramer's rule:

$$A = \frac{\begin{vmatrix} \Sigma y_i & \Sigma x_i \\ \Sigma x_i y_i & \Sigma x_i^2 \end{vmatrix}}{\begin{vmatrix} n & \Sigma x_i \\ \Sigma x_i & \Sigma x_i^2 \end{vmatrix}} \tag{15}$$

and

$$B = \frac{\begin{vmatrix} n & \Sigma y_i \\ \Sigma x_i & \Sigma x_i y_i \end{vmatrix}}{\begin{vmatrix} n & \Sigma x_i \\ \Sigma x_i & \Sigma x_i^2 \end{vmatrix}} \tag{16}$$

The corresponding equations we have to solve are:

$$A = \frac{\Sigma x_i^2 \Sigma y_i - \Sigma x_i \Sigma x_i y_i}{n\Sigma x_i^2 - \Sigma x_i \Sigma x_i} \tag{17}$$

and

$$B = \frac{n\Sigma x_i y_i - \Sigma x_i \Sigma y_i}{n\Sigma x_i^2 - \Sigma x_i \Sigma x_i} \tag{18}$$

The computer calculation of A and B is straightforward. The summation of x is obtained by summing the values of the array X. The summation of x^2 is obtained by squaring each value of X and then adding up the squares.

Equations 17 and 18 are commonly converted into an equivalent form by dividing the numerator and denominator by n.

$$A = \frac{(\Sigma x_i^2 \Sigma y_i - \Sigma x_i \Sigma x_i y_i)/n}{\Sigma x_i^2 - \Sigma x_i \Sigma x_i/n} \tag{19}$$

and

$$B = \frac{\Sigma x_i y_i - \Sigma x_i \Sigma y_i/n}{\Sigma x_i^2 - \Sigma x_i \Sigma x_i/n} \tag{20}$$

The denominators of Equations 19 and 20 appear in the formula for the standard deviation discussed in Chapter 2.

We have outlined the mathematics of least-squares curve-fitting and have derived formulas for finding the slope (B) and y-intercept (A) of a linear fitted curve. Now we are ready to add this feature to our program.

The Curve-Fitting Procedure

A program for fitting a straight line is shown in Figure 5.6. Make a copy of the previous program (Figure 5.4), then alter the subroutine starting at line 5000 so that it looks like the one in Figure 5.6.

```
10     REM Linear least—squares fit, Apr 19, 1981
11     REM
12     REM identifiers
14     REM      F2      FUDGE      fudge factor
15     REM      L3%     NLIN%      number of plot lines
16     REM      M1%     MAX%       maximum length
17     REM      N1%     NROW%      number of rows
18     REM      Y2      YCALC      calculated Y
19     REM end of identifiers
30     CLEAR 200
40     A$ = '' ###   ##.#   ###.##   ###.##''
50     REM
60     M1% = 35
70     DIM X(35), Y(35), Y2(35)
80     REM
90     PRINT
100    REM
110    GOSUB 500 : REM get the data
120    REM call to sorting routine goes here
130    REM
140    GOSUB 5000 : REM Perform curve fit
150    GOSUB 1000 : REM Print results
160    GOSUB 7000 : REM plot the data
170    GOTO  110 : REM next fit
```

Figure 5.6: Program to Generate a Least-Squares Fit

```
500    REM get the data
510    A = 2
520    B = 5
530    INPUT "Fudge"; F2
540    IF (F2 < 0) THEN 9999
550    INPUT "How many points"; N1%
560    FOR I% = 1 TO N1%
570        J% = N1% + 1 − I%
580        X(I%) = J%
590        Y(I%) = (A + B * J%) * (1 + (2 * RND(1) − 1) * F2)
600    NEXT I%
610    L3% = (N1% − 1) * 2 + 1
620    RETURN : REM from input routine
1000   REM print results
1010   PRINT "          X       Y     Y Calc"
1020   FOR I% = 1 TO N1%
1030       PRINT USING A$; I%; X(I%), Y(I%), Y2(I%)
1040   NEXT I%
1050   PRINT
1060   PRINT "Coefficients"
1070   PRINT "Intercept is "; A
1080   PRINT "     Slope is "; B
1090   PRINT
1100   REM
1110   RETURN : REM from printout
5000   REM Fit a straight line through y vs x, APR 19, 81
5010   REM
5012   REM identifiers
5014   REM     N1%    NROW%    number of rows
5015   REM     S3     SUMX     sum of x vector
5016   REM     S4     SUMX2    sum of x squared
5017   REM     S7     SUMY     sum of y vector
5018   REM     S8     SUMY2    sum of y squared
```

Figure 5.6: Program to Generate a Least-Squares Fit (cont.)

```
5019  REM     S9      SUMXY       sum of x times y
5020  REM     T2      SXX
5021  REM     T3      SXY
5022  REM     T4      SYY
5023  REM     Y2      YCALC       calculated Y
5024  REM end of identifiers
5030  S3 = 0 : S7 = 0
5040  S4 = 0 : S8 = 0
5050  S9 = 0
5060  FOR K% = 1 TO N1%
5070    X = X(K%)
5080    Y = Y(K%)
5090    S3 = S3 + X
5100    S7 = S7 + Y
5110    S9 = S9 + X * Y
5120    S4 = S4 + X * X
5130    S8 = S8 + Y * Y
5140  NEXT K%
5150  T2 = S4 − S3 * S3 / N1%
5160  T3 = S9 − S3 * S7 / N1%
5170  T4 = S8 − S7 * S7 / N1%
5180  B = T3 / T2
5190  A = ((S4 * S7 − S3 * S9)/ N1%) / T2
5240  FOR K% = 1 TO N1%
5250    Y2(K%) = A + B * X(K%)
5260  NEXT K%
5270  RETURN : REM from least squares fit
7000  REM Plot Y and Y2 as a function of X, Apr 14, 81
      (Continue with lines 7010 – 7980 of Figure 5.4.)

7990  RETURN : REM end of plot
9999  END
```

Figure 5.6: Program to Generate a Least-Squares Fit (cont.)

At the beginning of the new subroutine, the variables S3, S7, etc., which accumulate the needed sums, are set to zero and the **FOR** loop is used to calculated the desired sums. Notice that a change of variable is made at the beginning of the loop:

$$X = X(K\%)$$
$$Y = Y(K\%)$$

It generally takes longer to access an element of an array than to access a scalar value. Consequently, when the same value of an array is needed many times in a loop, it may be faster to define a scalar value and use that instead. On the other hand, some compilers incorporate an *optimizer* that automatically performs this task. In this case the transformation is unnecessary.

Running the Curve-Fitting Program

Run the new version. Give a fudge factor of 0.2 at first. The fitted line of asterisks should go neatly through the scattered plus symbols. Compare Figure 5.7, which shows the actual fit, with Figure 5.5, which gives the simulated fit. Next, make the fudge factor equal to zero. Only a single line of stars should be apparent now. Furthermore, the intercept should be equal to 2 and the slope should be equal to 5, the initial values.

THE CORRELATION COEFFICIENT

Although we now have a procedure for calculating the desired straight line, we are not finished yet. We can obtain the equation of a line fitting our experimental data. Then we can use this equation to predict a value for *y* from a given value of *x*. Under certain circumstances, however, our mathematically correct solution is useless.

Consider, for example, the set of data shown in Figure 5.8. Our curve-fitting program can find the equation of a straight line through the data, but the resulting line does not give us any additional information. That is, a knowledge of the behavior of *x* does not tell us anything about the behavior of *y*. There is no correlation between *x* and *y*.

The data shown in Figure 5.9 is another case where a knowledge of *x* is no help in predicting the behavior of *y*. Again, there is no correlation between *x* and *y*.

We need to obtain a quantitative measure of the correlation between *x* and *y*. We want to know how well we can predict the behavior of *y* if we know the behavior of *x*. The measure we need is the *correlation coefficient*.

```
Fudge? 0.2
How many points? 9
            X        Y      Y Calc
     1     9.0    42.21     42.95
     2     8.0    38.72     38.63
     3     7.0    34.22     34.30
     4     6.0    32.19     29.97
     5     5.0    22.23     25.64
     6     4.0    24.54     21.32
     7     3.0    16.98     16.99
     8     2.0    11.35     12.66
     9     1.0     8.36      8.33

Coefficients
Intercept is   4.00556
    Slope is   4.32771
```

Figure 5.7: A Least-Squares Fit to y vs x

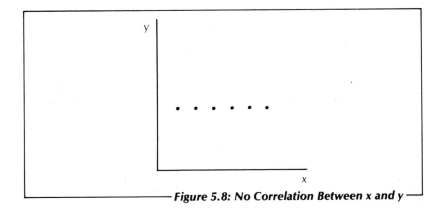

Figure 5.8: No Correlation Between x and y

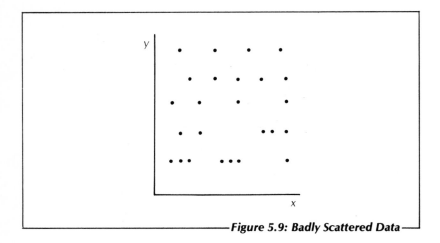

Figure 5.9: Badly Scattered Data

We saw in Chapter 2 that we could characterize a set of data by the mean and the standard deviation of the values about their own mean. We can also calculate the standard deviation of y about the fitted curve. This measure is termed the *standard error of the estimate (SEE)*. The correlation coefficient compares the standard deviation of y (about its own mean) to the standard deviation about the fitted curve.

The correlation coefficient is zero when there is no correlation. The data in Figures 5.8 and 5.9 are examples of this. On the other hand, the correlation coefficient approaches unity as the data approach a straight line. The correlation coefficient for the data given in Figure 5.7 is 0.99.

We will now incorporate calculations for the correlation coefficient into our program.

BASIC Subroutine for Calculating the Correlation Coefficient

With a few additional statements, our program can calculate the correlation coefficient and the standard errors of the coefficients (the intercept and slope). The standard errors are standard deviations for the coefficients. They can be used to determine confidence intervals for each coefficient.

Add the four lines:

```
5200 C3 = T3 / SQR(T2 * T4)
5210 T5 = SQR((S8 — A * S7 — B * S9) / (N1% —2))
5220 E3 = T5 / SQR(T2)
5230 E2 = E3 * SQR(S4 / N1%)
```

to the curve-fitting subroutine of your program. Then alter lines 1070-1100 so they look like:

1070 **PRINT** "Intercept is "; A; " Sigma is"; E2
1080 **PRINT** "   Slope is "; B; " Sigma is"; E3
1090 **PRINT**
1100 **PRINT** "Correlation coefficient is"; C3

With the correlation coefficient we have added a final test to our program—a way of quantifying the usefulness of the fitted curve. Let us look now at the finished program.

BASIC PROGRAM: LEAST-SQUARES CURVE FITTING FOR SIMULATED DATA

Run the new version. First set the fudge factor to a value of zero. The intercept will be 2 and the slope will be 5, as before. In addition, the sigmas for A and B will be zero and the correlation coefficient will be equal to one. The plot will show a straight line of stars.

Now try a fudge factor of 0.2. This will give sigmas greater than zero for the intercept and the slope. The correlation coefficient will be some-what less than unity. The entire source program is given in Figure 5.10.

```
10    REM Linear least—squares fit, Apr 19, 1981
11    REM
12    REM identifiers
14    REM     C3      CORREL      correlation coefficient
15    REM     E2      SIGMAA      error on A
17    REM     E3      SIGMAB      error on B
18    REM     F2      FUDGE       fudge factor
22    REM     L3%     NLIN%       number of plot lines
23    REM     M1%     MAX%        maximum length
24    REM     N1%     NROW%       number of rows
25    REM     Y2      YCALC       calculated Y
26    REM end of identifiers
30    CLEAR 200
```

Figure 5.10: The Complete Curve-Fitting Program

```
40    A$ = " ###   ##.#   ###.##   ###.##"
50    C$ = "  ##.####     ##.###^^^^"
60    M1% = 35
70    DIM X(35), Y(35), Y2(35)
80    REM
90    PRINT
100   REM
110   GOSUB 500 : REM get the data
120   REM call to sorting routine goes here
130   REM
140   GOSUB 5000 : REM Perform curve fit
150   GOSUB 1000 : REM Print results
160   GOSUB 7000 : REM plot the data
170   GOTO 110 : REM next fit
180   REM
500   REM get the data
510   A = 2
520   B = 5
530   INPUT "Fudge"; F2
540   IF (F2 < 0) THEN 9999
550   INPUT "How many points"; N1%
560   FOR I% = 1 TO N1%
570       J% = N1% + 1 - I%
580       X(I%) = J%
590       Y(I%) = (A + B * J%) * (1 + (2 * RND(1) - 1) * F2)
600   NEXT I%
610   L3% = (N1% - 1) * 2 + 1
620   RETURN : REM from input routine
1000  REM print results
1010  PRINT "        X        Y      Y Calc"
1020  FOR I% = 1 TO N1%
1030      PRINT USING A$; I%; X(I%), Y(I%), Y2(I%)
1040  NEXT I%
1050  PRINT
1060  PRINT "Coefficients"
```

Figure 5.10: The Complete Curve-Fitting Program (cont.)

```
1070    PRINT "Intercept is "; A; " Sigma is"; E2
1080    PRINT "      Slope is "; B; " Sigma is"; E3
1090    PRINT
1100    PRINT "Correlation coefficient is"; C3
1110    RETURN : REM from printout
5000    REM Fit a straight line through Y vs X, APR 19, 81
5010    REM
5012    REM identifiers
5014    REM     C3      CORREL      correlation coefficient
5015    REM     E2      SIGMAA      error on A
5016    REM     E3      SIGMAB      error on B
5017    REM     N1%     NROW%       number of rows
5018    REM     S3      SUMX        sum of x vector
5019    REM     S4      SUMX2       sum of x squared
5020    REM     S7      SUMY        sum of y vector
5021    REM     S8      SUMY2       sum of y squared
5022    REM     S9      SUMXY       sum of x times Y
5023    REM     T2      SXX
5024    REM     T3      SXY
5025    REM     T4      SYY
5026    REM     T5      SEE         stand. error of estimate
5027    REM     Y2      YCALC       calculated Y
5028    REM end of identifiers
5030    S3 = 0 : S7 = 0
5040    S4 = 0 : S8 = 0
5050    S9 = 0
5060    FOR K% = 1 TO N1%
5070        X = X(K%)
5080        Y = Y(K%)
5090        S3 = S3 + X
5100        S7 = S7 + Y
5110        S9 = S9 + X * Y
5120        S4 = S4 + X * X
5130        S8 = S8 + Y * Y
5140    NEXT K%
```

Figure 5.10: The Complete Curve-Fitting Program (cont.)

```
5150    T2 = S4 — S3 * S3 / N1%
5160    T3 = S9 — S3 * S7 / N1%
5170    T4 = S8 — S7 * S7 / N1%
5180    B = T3 / T2
5190    A = ((S4 * S7 — S3 * S9) / N1%) / T2
5200    C3 = T3 / SQR(T2 * T4)
5210    T5 = SQR((S8 — A * S7 — B * S9) / (N1% —2))
5220    E3 = T5 / SQR(T2)
5230    E2 = E3 * SQR(S4 / N1%)
5240    FOR K% = 1 TO N1%
5250       Y2(K%) = A + B * X(K%)
5260    NEXT K%
5270    RETURN : REM from least squares fit
7000    REM Plot Y and Y2 as a function of X, Apr 14, 81
        (Continue with lines 7010 – 7980 of Figure 5.4.)

7990    RETURN : REM end of plot
9999    END
```

Figure 5.10: The Complete Curve-Fitting Program (cont.)

SUMMARY

The modular development process that we have used for this program has allowed us to evaluate each subroutine as we wrote it. We began with the main program and added the subroutines step by step to plot the results, simulate data, compute the fitted curve, and supply the correlation coefficient. Significantly enough, our final curve-fitting routine replaced an earlier version that we wrote solely for the purpose of testing the plotting routine.

While we now have a linear curve-fitting program that works, it is not very useful. It can only fit data produced by a random number generator. Consequently, we will want to alter the input routine so that it can obtain actual data that is embedded in the program, or read from the keyboard or a disk file. We will delay this step, however, until we add a sorting routine in the next chapter.

EXERCISES

5-1: *The vapor pressure of water is:*

temp, C	press, torr
20	17.535
30	31.824
40	55.324
50	92.51
60	149.38
70	233.7
80	355.1
90	525.76
100	760.00

where the temperature is given in degrees Celsius and the pressure is in torr (mm. of mercury). Find the coefficients A and B to the equation:

$$\ln P = A + B/T$$

by performing a least-squares fit on the data. Be sure to convert the temperature to Kelvin by adding 273.15 to the Celsius value. Change the input routine to generate the x values as the reciprocal of the temperature in Kelvin:

570 X(I%) = 1.0/((I%−1) ∗ 10 + 20 + 273.15)

The y values can be read from a data statement, then converted to the logarithm:

580 **READ** Y(I%)
590 Y(I%) = LOG(Y(I%))
. . .
640 **DATA** 17.535, 31.824
. . .

Answer: A = 20.46, B = −5153

5-2: *The plotter routine given in Figure 5.4 prints the value of x at the left side of the page. The symbols + and ∗ are later printed to the right. This causes the plot symbols to be shifted to the right for large values of x. Incorporate the **TAB** function in the plotter routine so that the plot symbols always start at the same place regardless of the size of x.*

5-3: *Incorporate a range check in the plot routine given in Figure 5.4 so that values of x and y will be given in the E notation if they are very small or very large.*

5-4: *The plotter routine can only print one symbol in a particular location. Thus, if a plus symbol is coincident with a star symbol or another plus symbol, only one symbol is displayed. Alter the plot routine so that an M symbol is displayed whenever there is more than one element at a particular location.*

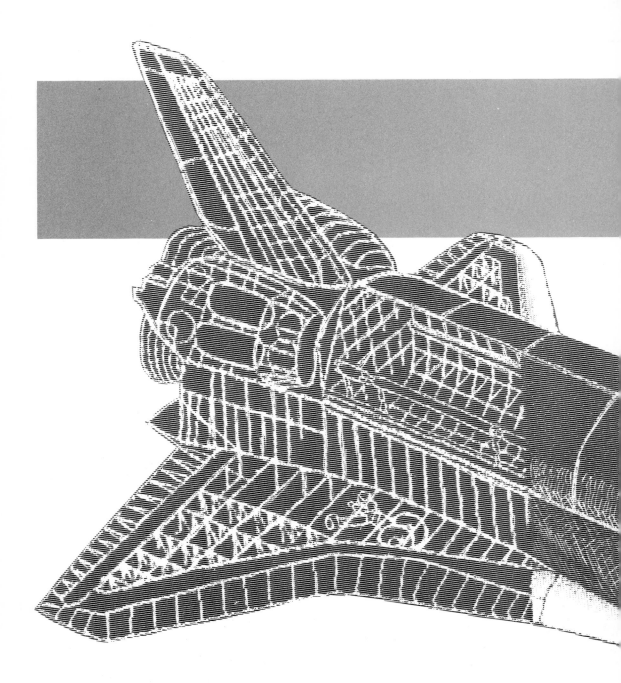

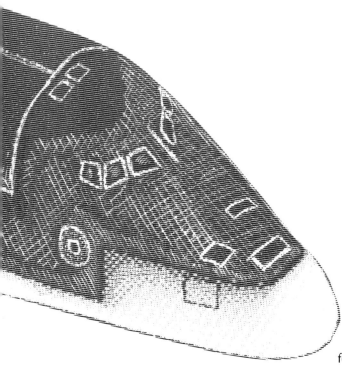

CHAPTER **6**

Sorting

INTRODUCTION

In this chapter we will develop several different sorting algorithms: two bubble sorts, a Shell sort and a nonrecursive quick sort. Then, we will incorporate one of these routines into the curve-fitting program developed in the previous chapter. First, let us discuss the rationale for including a sort routine in our program.

HANDLING EXPERIMENTAL DATA

The curve-fitting program we wrote in the previous chapter obtained its data from a random number generator. More realistic programs will obtain the data from the keyboard, or from a disk file. Alternately, the data can be embedded in the program itself.

In particular, we might want to fit experimental data that have not been acquired in numeric order. As an example, suppose that the thermal expansion of a material is to be investigated. Paired measurements of temperature and the corresponding total length could be taken. The experimental apparatus is brought to a certain temperature and the length is measured. Then the temperature is increased and the sample temperature and length are measured again. However, it would not be wise to continue the experiment in this fashion. The problem is that a length might be measured before the sample temperature became uniform, or before the sample reached the value of the temperature sensor. The resulting measured pairs would all contain errors in the same direction.

A better experimental technique would be to approach the desired temperature sometimes from below and sometimes from above. In this case, the experimental data might look like:

Temperature	Length
100	8.0
300	19.0
200	13.5
500	30.0
400	24.5

The curve-fitting program we developed in the last chapter could readily find a least-squares fit to this data. We could place the data directly in the input routine starting at line 500.

```
   . . .
510 N1% = 5
520 X(1) = 100 : Y(1) = 8.0
530 X(2) = 300 : Y(2) = 19.0
540 X(3) = 200 : Y(3) = 13.5
550 X(4) = 500 : Y(4) = 30.0
560 X(5) = 400 : Y(5) = 24.5
```

If we now run the program, however, we find that the plotting routine will give an incorrect rendering of the data. The problem is that the array of independent variables, X, must be arranged either in increasing or in decreasing order. That is, the data must be sorted. The plotting subroutine worked properly in the previous chapter because the X array was generated in decreasing order. Notice that we are not concerned about the ordering of the dependent variable Y.

We will begin our discussion of sorting routines with the easiest one to program—the bubble sort.

A BUBBLE SORT

To sort a collection of items is to arrange them in increasing or decreasing order. The items might be elements of an array of real numbers, or they might be a group of alphabetic and numeric characters called records. There are many different sorting algorithms. Some are very fast, others are very slow. Some are faster with nearly sorted data and some are slower under these conditions. Some require additional working space, others utilize only the space occupied by the original data.

The first sorting algorithm we will consider is known as the *bubble sort*. This routine is the easiest to understand and the easiest to program. Unfortunately, it is also the slowest. For sorting lists that contain fewer than a dozen or so items, however, the difference in speed is unimportant.

With the bubble sort, each element is compared to all of the remaining, unsorted elements. If a particular pair is found to be out of order, the two elements are interchanged. There are two loops, one nested inside the other. The outer loop runs from 1 to one less than the length of the array. The inner loop runs from one larger than the outer loop up to the length of the array. With this algorithm, the smaller items "float" to the top of the array during the sorting process. This is of course the origin of the name "bubble sort".

We will incorporate this first sorting routine into a complete program designed to test the relative efficiency of different sorting routines under different conditions. The sorting routine itself will begin at line 3000. We will also begin the other sort routines in this chapter at the same line number. This will make it easy to add any of the sorting routines to other programs.

BASIC PROGRAM: THE BUBBLE SORT

The program shown in Figure 6.1 contains a driver routine in addition to the sorting program. When the program is executed, it asks the user for the number of items to be sorted. A random number generator is then called. It generates the desired number of elements in the range of zero to 100. The RND function may need to be changed to RND(0) or RND. The original set of numbers is printed on the console. The expression CHR$(7), at lines 140, 160, 200, 260 and 280, sounds the console bell at the beginning and at the ending of each sorting step. This will allow a comparison to be made with the other sorting routines to be presented in this chapter. If you are not using an ASCII terminal, then this expression will have to be changed or removed.

At the end of the sorting step, the sorted array is printed on the console. The word "random" is also printed. The console bell sounds a third time, and the sorting routine is called again. This time, however, the sort is performed on an array that is already sorted. At the end of this second sorting step, the console bell sounds a fourth time and the sorted array is printed out again. The word "sorted" is also printed. For the third phase of the test program, an array of numbers in reverse order is generated. The sorting routine is then called for the third time. At the end of the step, the sorted array is printed along with the word "reversed."

Type up the program given in Figure 6.1 and run it. Try to find a length that requires several minutes for sorting. Record the time needed for sorting each of the three arrangements of data. Then a comparison can be made with the other sorting routines given later in this chapter. If you are not using a video terminal, you may want to disable the printing of the arrays. Replace line 600 with a **RETURN** statement.

The programs in this chapter require function RND for the generation of random numbers. If your BASIC does not contain this function, you should include the random number generator given in Figure 5.2 of Chapter 5.

```
10   REM test speed of sorting routine, Apr 21, 81
11   REM
12   REM identifiers
14   REM      H1      HOLD
15   REM      M1%     MAX%      maximum length
16   REM      N1%     NROW%     number of rows
17   REM end of identifiers
30   REM
40   M1% = 500
50   DIM X(500)
```

Figure 6.1: A Bubble-Sort Subroutine and Driving Program

```
  60   REM
  70   INPUT " How many points"; N1%
  80   IF (N1% < 5) THEN 9999
  90   IF (N1% > M1%) THEN 70
 100   FOR I% = 1 TO N1%
 110      X(I%) = RND(1) : REM May need to be RND(0) or RND
 120   NEXT I%
 130   GOSUB 600 : REM print original array
 140   PRINT CHR$(7)
 150   GOSUB 3000 : REM sort random numbers
 160   PRINT CHR$(7)
 170   GOSUB 600 : REM print sorted array
 180   PRINT " random"; CHR$(7)
 190   GOSUB 3000 : REM sort sorted numbers
 200   PRINT CHR$(7)
 210   GOSUB 600
 220   PRINT " sorted"
 230   FOR I% = 1 TO N1%
 240      X(I%) = N1% + 1 - I%
 250   NEXT I%
 260   PRINT CHR$(7)
 270   GOSUB 3000 : REM sort reversed data
 280   PRINT CHR$(7)
 290   GOSUB 600
 300   PRINT " reversed"
 310   GOTO 70
 600   REM print array X
 610   PRINT
 620   FOR I% = 1 TO N1%
 630      PRINT X(I%),
 640   NEXT I%
 650   RETURN : REM from output routine
3000   REM bubble sort routine
```

Figure 6.1: A Bubble-Sort Subroutine and Driving Program (cont.)

```
3010    FOR I% = 1 TO N1% − 1
3020      FOR J% = I% + 1 TO N1%
3030        IF (X(I%) < X(J%)) THEN 3070
3040        H1 = X(I%)
3050        X(I%) = X(J%)
3060        X(J%) = H1
3070      NEXT J%
3080    NEXT I%
3090    RETURN
9999    END
```

Figure 6.1: A Bubble-Sort Subroutine and Driving Program (cont.)

We will now develop a slightly improved bubble sort routine. The reason for this improvement will become clearer as the chapter progresses.

Adding a Swap Function

Later in this chapter, we will incorporate a sorting routine into the curve-fitting program from the previous chapter. At that time, we will want to interchange a value of a second array (Y) whenever we interchange the corresponding value of the first array (X). The middle part of the sort routine might then become:

```
H1 = X(I%)
X(I%) = X(J%)
X(J%) = H1
H1 = Y(I%)
Y(I%) = Y(J%)
Y(J%) = H1
```

In this case, a pair of elements of one array is first interchanged, and then the corresponding elements in the other array are swapped. As in Figure 6.1, the interchange operation requires a third variable called H1.

While it is not a standard BASIC feature, the SWAP function is incorporated into some BASICs. If you have such a feature, it should be utilized in the sorting routine. With this approach, each interchange requires only a single instruction. We will explore the advantages of this approach in our second version of the bubble sort routine.

BASIC PROGRAM: BUBBLE SORT WITH SWAP FUNCTION

In the previous program we used a temporary variable called H1 for the interchange of two values in the array X. This interchange can be performed more easily if there is a BASIC function called SWAP. The expression:

SWAP A, B

will swap the variables A and B. There are two advantages to using the SWAP function. The source code is easier to comprehend and the resulting program will run faster. The speed improvement is especially noticeable when arrays of strings are being sorted. The reason for the faster operation is that the SWAP function interchanges the pointers to the variables, rather than the variables themselves.

An additional advantage to the swap function will be apparent when we add the sort routine to the curve-fitting program of the previous chapter. In this case, when we interchange two values of the array X we also have to interchange the two corresponding values of array Y.

SWAP X(I%), X(J%)
SWAP Y(I%), Y(J%)

The sorting routine shown in Figure 6.2 demonstrates the use of the SWAP function. The algorithm is a variation of the bubble sort. The new version will run much faster than the first one if the original array is already ordered. Otherwise, it will run more slowly.

```
  10   REM test speed of sorting routine
       (Continue with lines 20 – 640 of Figure 6.1.)
 650   RETURN : REM from output routine
3000   REM bubble sort routine
3011   REM identifiers
3013   REM      N1%    NROW%    number of rows
3014   REM      N5%    NSWAP%   swap flag
3015   REM end of identifiers
3016   REM
3020   N5% = 1
3030   REM
3040   N5% = 0
```

Figure 6.2: A Variation of the Bubble Sort, Incorporating Function SWAP

```
3050    FOR J% = 1 TO N1% - 1
3060        IF (X(J%)< = X(J%+1)) THEN 3090
3070            SWAP X(J%), X(J%+1)
3080            N5% = 1
3090    NEXT J%
3100    IF (N5% = 1) THEN 3040
3110    RETURN : REM from bubble sort
9999    END
```

Figure 6.2:
A Variation of the Bubble Sort, Incorporating Function SWAP (cont.)

Running the New Version

Incorporate the new version of subroutine SORT in place of the original. Make a copy first, then alter the copy so it looks like the listing in Figure 6.2. Execute the new version and compare the sorting times to the first version.

Next we will study a more sophisticated and slightly more difficult sorting routine.

A SHELL SORT

The major disadvantage with the bubble sort method is that it frequently makes more comparisons and more interchanges than are necessary. An item may be moved from one end of the array to the other. Then, a little later, it might be moved nearly back to where it started.

The *Shell-Metzner* sort (or *Shell* sort) is generally more efficient than the bubble sort. Comparisons are initially made over long distances. The first item in the array is compared to one in the middle, rather than to the one right next door. For short lists of fewer than a dozen items, the Shell sort and bubble sort are comparable in speed. But, as the length of the list increases, the speed of the Shell sort becomes apparent. The Shell sort is noticeably faster than the bubble sort when the number of items exceeds about fifty.

We will be able to quantify our comparison of the bubble and Shell sorts by comparing the sorting times.

BASIC PROGRAM: THE SHELL-METZNER SORT

Copy one of the previous sorting programs and alter the new copy so that it looks like the listing in Figure 6.3. Run the new program and compare the sorting times to the bubble sort. You should find that the Shell sort runs much faster. You will also notice that the Shell sort, like the bubble sort, runs much faster with sorted data than with unsorted data.

In the discussion of our last and most complex sorting routine we will present a nonrecursive version of the quick sort routine.

```
10     REM test speed of sorting routine
       (Continue with lines 20 – 640 of Figure 6.1.)

650    RETURN : REM from output routine
3000   REM Shell — Metzner sort routine
3011   REM identifiers
3013   REM      H1        HOLD
3014   REM      J6%       JUMP%
3015   REM      N1%       NROW%        number of rows
3016   REM end of identifiers
3020   REM
3030   J6% = N1%
3040   J6% = INT(J6% / 2)
3050   IF (J6% = 0) THEN 3180
3060   J2% = N1% − J6%
3070   FOR J% = 1 TO J2%
3080      I% = J%
3090      J3% = I% + J6%
3100      IF (X(I%) < = X(J3%)) THEN 3160
3110      H1 = X(I%)
3120      X(I%) = X(J3%)
3130      X(J3%) = H1
3140      I% = I% − J6%
3150      IF (I% > 0) THEN 3090
3160   NEXT J%
3170   GOTO 3040
3180   RETURN : REM from Shell sort
9999   END
```

Figure 6.3: A Shell Sort Program

THE QUICK SORT

We saw that the bubble sort is easy to program and easy to understand. The Shell sort is a bit more complicated, but it can sort much more quickly. The third sorting algorithm we will consider is known as *quick* sort. It is even more complicated than the previous algorithms. It is generally faster than the bubble sort or the Shell sort, although if the original data are already sorted, or nearly sorted, the Shell sort can be much faster. A quick sort routine takes almost as long to run on a sorted array as on an unsorted one.

BASIC PROGRAM: A NONRECURSIVE QUICK SORT

The bubble sort begins by comparing elements that are side by side. The Shell sort begins by comparing the initial element to one at the middle of the array. The quick sort begins by comparing the elements at the opposite ends of the array. Thus, the initial interchanges for quick sort can be made over large distances.

An array to be sorted is repeatedly partitioned into two smaller and smaller parts. The elements are rearranged so that all the elements on the one side are smaller than all those on the other. The two new sections are then divided into two subsections and the process is repeated. Partitioning continues until there are many sets containing one element each, at which point the array is sorted.

The quick sort algorithm is usually written to operate recursively; that is, one portion of the code calls itself. Recursion can be readily implemented in certain computer languages. However, in standard BASIC all variables are global. Consequently, recursive routines can only be implemented with difficulty. A nonrecursive version of quick sort is given in Figure 6.4. This algorithm simulates the recursive calls by saving the partition pointers in a stack. The left and right pointers are stored in integer arrays L5% and R4%, respectively. These variables must be included in the dimension statement at the beginning of the main program. If sophisticated features such as long variable names, SWAP, and **WHILE/WEND** are available, the program can be written in a much more readable style.

The pivot element is initially chosen to be the last element in the array. This is a poor choice if the array is already arranged in increasing or decreasing order. Consequently, this pivot element is compared with two other elements: the first element of the array and the element located at the center of the array. The element that is the median of these three, neither the largest nor the smallest, is chosen as the pivot. This change will greatly speed up the sorting process when the array or subarray is already ordered in one direction or the other.

One feature of this version is that at each partitioning, the smaller of the two subsets is sorted before the larger. This minimizes the space needed for storage of the unsorted indices.

You will want to keep versions of both the Shell sort and quick sort. The latter will generally be better, but if you frequently need to sort data that are almost sorted, you may find that Shell sort is more suitable.

In the next section we will develop a program for sorting data stored on disk.

```
 10    REM test speed of quick sort routine, Apr 21, 81
 11    REM
 12    REM identifiers
 14    REM      H1       HOLD
 15    REM      M1%      MAX%        maximum length
 16    REM      N1%      NROW%       number of rows
 17    REM end of identifiers
 30    REM
 40    M1% = 500
 50    DIM X(500), L5%(20), R4%(20)
 60    REM
 70    INPUT " How  many  points"; N1%
 80    IF (N1% < 5) THEN 9999
 90    IF (N1% > M1%) THEN 70
100      FOR I% = 1 TO N1%
110        X(I%) = RND(1) : REM May need to be RND(0) or RND
120    NEXT I%
130    GOSUB 600 : REM print original array
140    PRINT CHR$(7)
150    GOSUB 3000 : REM sort random numbers
160    PRINT CHR$(7)
170    GOSUB 600 : REM print sorted array
180    PRINT " random"; CHR$(7)
190    GOSUB 3000 : REM sort sorted numbers
200    PRINT CHR$(7)
210    GOSUB 600
220    PRINT " sorted"
230    FOR I% = 1 TO N1%
240      X(I%) = N1% + 1 − I%
250    NEXT I%
```

Figure 6.4: A Nonrecursive Quick Sort

```
260     PRINT CHR$(7)
270     GOSUB 3000 : REM sort reversed data
280     PRINT CHR$(7)
290     GOSUB 600
300     PRINT '' reversed''
310     GOTO 70
600     REM print array X
610     PRINT
620     FOR I% = 1 TO N1%
630        PRINT X(I%),
640     NEXT I%
650     RETURN : REM from output routine
3000    REM nonrecursive quick sort, Apr 15, 81
3010    REM
3011    REM identifiers
3013    REM      H1       HOLD
3014    REM      L5%      LEFT%        left pointer
3015    REM      M4%      MID%         middle pointer
3016    REM      N1%      NROW%        number of rows
3017    REM      P1       PIVOT        pivot element
3018    REM      P2%      SP%          stack pointer
3019    REM      R4%      RIGHT%       right pointer
3020    REM end of identifiers
3030    REM Declare DIM L5%(20), R4%(20) in main program
3040    REM in addition to vector X
3050    L5%(1) = 1
3060    R4%(1) = N1%
3070    P2% = 1
3080    IF (L5%(P2%) < R4%(P2%)) THEN 3110
3090        P2% = P2% − 1
3100        GOTO 3540
3110    I% = L5%(P2%)
3120    J% = R4%(P2%)
3130    P1 = X(J%)
3140    M4% = (I% + J%) / 2
```

Figure 6.4: A Nonrecursive Quick Sort (cont.)

```
3150    IF (J% − I% < 6) THEN 3280
3160    IF ((P1 > X(I%)) AND (P1 < X(M4%))) THEN 3280
3170    IF ((P1 < X(I%)) AND (P1 > X(M4%))) THEN 3280
3180    IF ((X(I%) < X(M4%)) AND (X(I%) > P1)) THEN 3240
3190    IF ((X(I%) > X(M4%)) AND (X(I%) < P1)) THEN 3240
3200    H1 = X(M4%)
3210    X(M4%) = X(J%)
3220    X(J%) = H1
3230    GOTO 3270
3240    H1 = X(I%)
3250    X(I%) = X(J%)
3260    X(J%) = H1
3270    P1 = X(J%)
3280    IF (I% >= J%) THEN 3410
3290      IF (X(I%) >= P1) THEN 3320
3300        I% = I% + 1
3310        GOTO 3290
3320      J% = J% − 1
3330      IF (NOT((I% <J%) AND (P1 <X(J%)))) THEN 3360
3340        J% = J% − 1
3350        GOTO 3330
3360      IF (I% >= J%) THEN 3400
3370        H1 = X(I%)
3380        X(I%) = X(J%)
3390        X(J%) = H1
3400    GOTO 3280
3410    J% = R4%(P2%)
3420    H1 = X(I%)
3430    X(I%) = X(J%)
3440    X(J%) = H1
3450    IF ( I% − L5%(P2%) >= R4%(P2%) − I%) THEN 3500
3460        L5%(P2% + 1) = L5%(P2%) : REM stack shorter first
3470        R4%(P2% + 1) = I% − 1
3480        L5%(P2%) = I% + 1
3490        GOTO 3530
```

Figure 6.4: A Nonrecursive Quick Sort (cont.)

```
3500    L5%(P2%+1) = I% + 1 : REM ELSE
3510    R4%(P2%+1) = R4%(P2%)
3520    R4%(P2%) = I% - 1 : REM end ELSE
3530    P2% = P2%+1 : REM push stack
3540    IF (P2% > 0) THEN 3080
3550    RETURN : REM from quick sort
9999    END
```

Figure 6.4: A Nonrecursive Quick Sort (cont.)

SORTING DISK DATA

One of the tasks that BASIC is ideally suited for is the handling of strings. A collection of records, which might represent names, addresses, etc., can be established as a disk file. The entire file can be addressed as a one-dimensional string array called A$. Each line or record can contain one name, address, etc. Thus, the first line, containing a name and address, is designated as A$(1), the second line is A$(2), etc. Furthermore, it is not necessary to block the records; that is, it is not necessary to make each record the same length.

The BASIC program given in Figure 6.5 can be used to sort a disk file of string records. Unfortunately, the commands that are used to read and write disk files have not been standardized. The dialect of disk commands used by Microsoft BASIC has been chosen in this case. Other powerful features, such as **WHILE/WEND**, integer division, long variable names, and the SWAP function, are also illustrated in this program.

```
10     REM Sort ASCII disk file
20     REM
30     MAX% = 550
40     DIM A$(500)
50     PRINT "Shell sort for string records"
60     PRINT : INPUT "File to be sorted"; OFIL$
70     A2$ = OFIL$
80     GOSUB 3700
90     OFIL$ = A2$
100    INPUT "New—file name"; NFIL$
110    A2$ = NFIL$
```

Figure 6.5: A Microsoft BASIC Program for Sorting String Disk Files

```
120     GOSUB 3700
130     NFIL$ = A2$
140     OPEN ''I'', #2, OFIL$
150     I% = 0
160     IF EOF(2) THEN 230
170     I% = I% + 1
180     IF I% > MAX% THEN 210
190     LINE INPUT #2, A$(I%)
200     GOTO 160
210     PRINT ''Lines exceed dimension of ''; MAX%
220     GOTO 9999
230     NROW% = I%
240     GOSUB 3000 : REM sort
250     OPEN ''O'', #1, NFIL$
260     FOR I% = 1 TO NROW%
270     PRINT #1, A$(I%)
280     NEXT I%
290     CLOSE #1 : CLOSE #2
300     FOR I% = 1 TO NROW%
310     PRINT A$(I%)
320     NEXT I%
330     GOTO 60
3000    REM Shell—Metzner sort routine
3010    REM
3020    JUMP% = NROW%
3030    WHILE JUMP% <> 0
3040       JUMP% = JUMP% \ 2
3050       J2% = NROW% − JUMP%
3060       J% = 1
3070       WHILE J% <= J2%
3080          I% = J%
3090          WHILE I% > 0
3100             J3% = I% + JUMP%
3110             IF (A$(I%) <= A$(J3%)) THEN 3150
```

Figure 6.5:
Microsoft BASIC Program for Sorting String Disk Files (cont.)

```
3120            SWAP A$(I%), A$(J3%)
3130             I% = I% − JUMP%
3140         WEND
3150         J% = J% + 1
3160       WEND
3170     WEND
3180   RETURN : REM from Shell sort
3700   REM convert lower case to upper
3710   L2% = LEN(A2$)
3720   A3$ = ""
3730   FOR I% = 1 TO L2%
3740      A4$ = MID$(A2$, I%, 1)
3750      IF A4$ > "Z" THEN A4$ = CHR$(ASC(A4$) AND 95)
3760      A3$ = A3$ + A4$
3770   NEXT I%
3780   A2$ = A3$
3790   RETURN
9999   END
```

Figure 6.5:
Microsoft BASIC Program for Sorting String Disk Files (cont.)

The sorting program demonstrates several particular features. When the program requests the name of the original unsorted file, the name can be entered in either upper or lower case letters. The program converts lower case letters to upper case in the subroutine starting at line 3700. This is accomplished by performing a logical AND with the value 95.

The file is read into the array A$. The records are sorted in increasing order, then the resulting sorted array is written to a new disk file. The name to be assigned to the new file is also requested from the user.

Several other features are peculiar to Microsoft BASIC. These include the use of the SWAP function, the **WHILE/WEND** construction and integer division at line 3040. The program can be easily altered so that it will arrange the data in reverse order. Change the $<$ = pair in line 3110 to a $>$ = pair.

Finally, let us return to our curve-fitting program, which is what originally led us into this discussion of sorting routines.

INCORPORATING SORT INTO THE CURVE-FITTING PROGRAM

The next step is to incorporate one of the sorting routines into the curve-fitting program we wrote in the previous chapter. As before, keep a working copy of the prior version. Then you will have something to go back to if the new version becomes hopelessly mixed up. Two changes will have to be made when the sorting routine is incorporated into the curve-fitting program. A call to the sorting subroutine is inserted at line 120 of the original program. (A remark was placed here for this purpose.) Additional instructions are needed to interchange elements of array Y when the corresponding elements of X are interchanged. For the Shell sort routine, add the lines:

```
3132 H1 = Y(I%)
3134 Y(I%) = Y(J3%)
3136 Y(J3%) = H1
```

SUMMARY

We have seen variations of three common sort routines: the bubble sort, the Shell sort and the quick sort. All of these sorting routines are written to operate on real numbers. Keep in mind that these procedures can easily be altered to sort integers, characters, or strings of characters.

Now that our curve-fitting program can handle real experimental data, we are ready to apply the program to some more complex equations. We will do this in Chapter 7.

EXERCISES

6-1: *Incorporate a direction flag, called F7%, into one of the sorting routines. Preset the flag to zero so that sorting will occur in the usual way. Ask the user if the data are to be sorted in reverse order. If the answer is positive, change the flag. In the Shell sort routine, for example, there might be the pair of expressions:*

IF ((F7% = 0) **AND** (X(I%) <= X(J3%))) **THEN** 3160
IF ((F7% <> 0) **AND** (X(I%) >= X(J3%))) **THEN** 3160

6-2: *Change one of the sorting routines so it will sort strings obtained from a set of **DATA** statements.*

6-3: *Change one of the sorting routines so it will sort strings obtained from the console. Use the LINE **INPUT** command if your BASIC has one.*

6-4: *Alter the program from Problem 6-2 so that the data are sorted on an interior field. Make columns 11 through 20 of the data correspond to a city. Then use these columns as the basis for sorting. The sorting criterion will be based on an expression such as:*

IF (MID$(X%(I), 11, 20) . . .

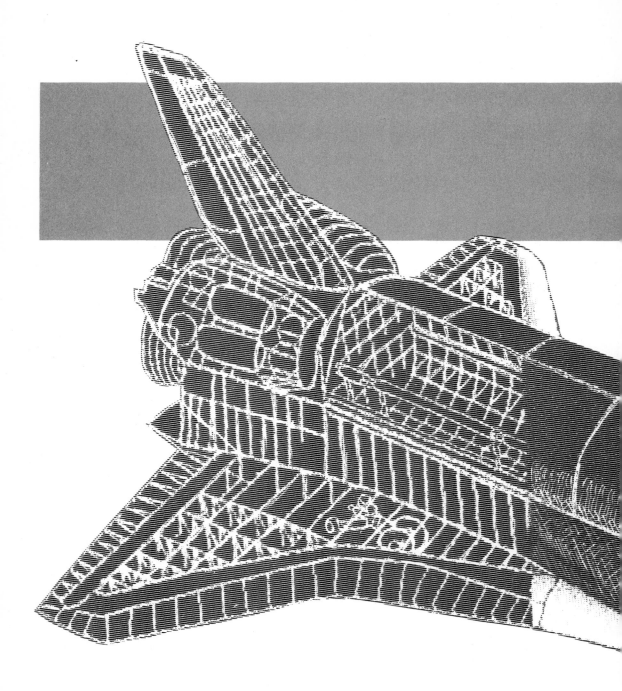

CHAPTER 7

General Least-Squares Curve Fitting

INTRODUCTION

In this chapter we will develop several different least-squares curve-fitting programs. Our goal will be to *generalize* the previous curve-fitting program so that it will be a more useful and realisitc tool for a wider variety of experimental situations. Up to now we have been limited to the equation of a straight line. Our first program in this chapter will implement a parabolic curve—that is, a second-order polynomial equation. From that point we will develop a method that will handle both higher-order polynomials and nonpolynomial equations. The only restriction will be that the unknown coefficients must be linear.

The key to our approach will be the use of a data vector and a data matrix. The method will subsequently allow us to input the *order* of a polynomial equation from the keyboard once the program is running. Finally, we will implement curve-fitting programs for some real experimental data. The equations of the fitted data will include those used for heat capacity and vapor pressure. In addition, in our program for a three-variable equation of state, we will experiment with solutions for a nonlinear coefficient.

A PARABOLIC CURVE FIT

A least-squares curve-fitting program was developed in Chapter 5. This program was used to calculate the coefficients A and B for the expression:

$$y = A + Bx \tag{1}$$

While Equation 1 is the most commonly used curve-fitting equation, there are times when a different equation is required. Therefore, in this chapter, we will extend the least-squares method to include other commonly used expressions.

We will place one restriction on the form of the curve-fitting equation. It must be linear in the unknown coefficients. Thus, we can consider equations such as:

$$y = A + Bx + Cx^2$$

$$y = A + \frac{B}{x} + Cz$$

$$\ln y = \frac{Ax}{z} + B\,e^x$$

because the coefficients A, B, and C are linear. But we will not consider an equation such as:

$$y = A + B\,e^{Cx}$$

since the coefficient C is not linear.

The development of a curve-fitting program for any of the above equations is similar to the approach we used in Chapter 5. Consider, for example, the parabolic equation, which is a second-order polynomial:

$$y = A + Bx + Cx^2 \tag{2}$$

We define the residuals to be:

$$r = A + Bx + Cx^2 - y$$

The residuals are squared, and then summed. As in Chapter 5, we wish to minimize this quantity. We thus take the derivative with respect to each variable (A, B, and C in this case), and set the resulting equations to zero. For the parabolic curve fit there will be three equations, one for each of the three variables. The resulting equations are:

$$An + B\Sigma x + C\Sigma x^2 = \Sigma y \tag{3}$$

$$A\Sigma x + B\Sigma x^2 + C\Sigma x^3 = \Sigma xy \tag{4}$$

$$A\Sigma x^2 + B\Sigma x^3 + C\Sigma x^4 = \Sigma x^2 y \tag{5}$$

BASIC PROGRAM: LEAST-SQUARES CURVE FIT FOR A PARABOLA

The solution to the parabolic fit is obtained by solving these three equations simultaneously. The program shown in Figure 7.1 finds the solution to these equations by using Cramer's rule. The determinants of four 3-by-3 matrices must be solved with this approach. The subroutine we will use for this purpose was developed in Chapter 4 (Figure 4.3). It begins at line 5280.

In the programs of the previous chapters, the coefficients were represented by the variables A, B, and C. In this chapter, however, we will use the array C1 for this purpose. Thus, the constant term, *A*, now corresponds to the first element of the vector, C1(1). Similarly, the coefficients *B* and *C* correspond to C1(2) and C1(3) respectively.

10	**REM** Parabolic fit By Cramer's rule, Apr 19, 1981			
11	**REM** identifiers			
13	**REM**	C1	COEF	solution vector
14	**REM**	C3	CORREL	correlation coefficient
15	**REM**	D3	DETERM	determinant
16	**REM**	E1%	ERMES%	error flag
17	**REM**	L3%	NLIN%	number of plot lines
18	**REM**	M1%	MAX%	maximum length
19	**REM**	N1%	NROW%	number of rows
20	**REM**	N2%	NCOL%	number of columns
21	**REM**	R5	RES	residual
22	**REM**	S3	SUMX	sum of x
23	**REM**	S6	SUM	
24	**REM**	S7	SUMY	sum of y
25	**REM**	S9	SUMXY	sum x times y
26	**REM**	T6	SRS	sum residuals squared
27	**REM**	T7	SUM2Y	sum x*x*y
28	**REM**	T8	SUMX3	sum x cubed
29	**REM**	T9	SUMX4	sum x fourth
30	**REM**	Y2	YCALC	calculated y
31	**REM** end of identifiers			
35	**CLEAR** 200			

Figure 7.1: A Parabolic Least-Squares Fit

```
 40   A$ = " ###   ##.#   ###.##   ###.##"
 50   C$ = "  ##.####      ##.###^^^^"
 60   M1% = 35
 70   DIM Z(3), A(3,3), C1(3), Y(35), U(35,3)
 80   DIM W(3,1), B(3,3), X(35), Y1(35)
 90   DIM Y2(35)
100   REM
110   PRINT
120   PRINT " Parabolic least—squares fit by Cramer's rule"
130   REM
140   GOSUB 500 : REM get the data
150   REM call to sorting routine goes here
160   REM
170   GOSUB 4000 : REM square up the matrix
180   GOSUB 5000 : REM Cramer's rule
190   GOSUB 1000 : REM Print results
200   GOSUB 7000 : REM plot data
210   GOTO 9999 : REM done
500   REM get the data
510   N1% = 9
520   N2% = 3
530   FOR I% = 1 TO N1%
540      X(I%) = I%
550      READ Y(I%)
560   NEXT I%
570   L3% = (N1% — 1) * 2 + 1
580   REM Y data
590   DATA 2.07, 8.6, 14.42, 15.8
600   DATA 18.92, 17.96, 12.98, 6.45, 0.27
610   RETURN : REM from input routine
1000  REM Calculate residuals and print results
1010  T6 = 0
1020  FOR I% = 1 TO N1%
1030     X = X(I%)
```

Figure 7.1: A Parabolic Least-Squares Fit (cont.)

```
1040        Y2(I%) = C1(1) + C1(2) * X + C1(3) * X * X
1050        R5 = Y(I%) — Y2(I%)
1060        T6 = T6 + R5 * R5
1070    NEXT I%
1080    REM S7 and S8 are from subroutine 4000
1090    C3 = SQR(1 — T6 / (S8 — S7 * S7 / N1%))
1100    PRINT "          X        Y       Y Calc"
1110    FOR I% = 1 TO N1%
1120        PRINT USING A$; I%; X(I%), Y(I%), Y2(I%)
1130    NEXT I%
1140    PRINT
1150    PRINT "Coefficients"
1160    PRINT USING C$; C1(1);
1170    PRINT "  Constant  term"
1180    FOR I% = 2 TO N2%
1190        PRINT USING C$; C1(I%)
1200    NEXT I%
1210    PRINT
1220    PRINT "Correlation  coefficient  is"; C3
1230    RETURN : REM from printout
4000    REM Form square matrix A and vector Z
4010    S3 = 0 : S7 = 0
4020    S9 = 0 : S4 = 0
4030    S8 = 0 : T8 = 0
4040    T9 = 0 : T7 = 0
4050    REM
4060    FOR K% = 1 TO N1%
4070        X = X(K%)
4080        Y = Y(K%)
4090        X2 = X * X
4100        S3 = S3 + X
4110        S7 = S7 + Y
4120        S9 = S9 + X * Y
4130        S4 = S4 + X2
```

Figure 7.1: A Parabolic Least-Squares Fit (cont.)

```
4140        S8 = S8 + Y * Y
4150        T8 = T8 + X * X2
4160        T9 = T9 + X2 * X2
4170        T7 = T7 + X2 * Y
4180    NEXT K%
4190    A(1,1) = N1%
4200    A(2,1) = S3
4210    A(1,2) = S3
4220    A(3,1) = S4
4230    A(1,3) = S4
4240    A(2,2) = S4
4250    A(3,2) = T8
4260    A(2,3) = T8
4270    A(3,3) = T9
4280    Z(1) = S7
4290    Z(2) = S9
4300    Z(3) = T7
4310    RETURN : REM from matrix setup
5000    REM Solution of a 3—by—3 matrix by Cramer's rule
5010    REM Apr 15, 81
5011    REM identifiers
5013    REM     A       A           coefficient matrix
5014    REM     B       B           work array
5015    REM     C1      COEF        solution vector
5016    REM     D3      DETERM      determinant
5017    REM     E1%     ERMES%      error flag
5018    REM     N2%     NCOL%       number of columns
5019    REM     S6      SUM
5020    REM     Z       Z           constant vector
5021    REM end of identifiers
5070    REM
5080    FOR I% = 1 TO N2%
5090        FOR J% = 1 TO N2%
5100            B(I%,J%) = A(I%,J%)
```

Figure 7.1: A Parabolic Least-Squares Fit (cont.)

```
5110        NEXT J%
5120      NEXT I%
5130      GOSUB 5280 : REM find the determinant
5140      D3 = S6
5150      IF (D3 = 0) THEN 5250
5160      FOR J% = 1 TO N2%
5170        FOR I% = 1 TO N2%
5180          B(I%,J%) = Z(I%)
5190          IF (J% > 1) THEN B(I%,J%−1) = A(I%,J%−1)
5200        NEXT I%
5210        GOSUB 5280 : REM find the determinant
5220        C1(J%) = S6 / D3
5230      NEXT J%
5240      RETURN : REM normal return from Cramer's rule
5250      E1% = 1
5260      PRINT "ERROR−matrix  singular  "
5270      RETURN : REM from Cramer's rule
5280      REM Find the determinant of a 3−by−3 matrix
5290      REM
5300      S6 = B(1,1) * (B(2,2) * B(3,3) − B(3,2) * B(2,3))
5310      S6 = S6 − B(1,2) * (B(2,1) * B(3,3) − B(3,1) * B(2,3))
5320      S6 = S6 + B(1,3) * (B(2,1) * B(3,2) − B(3,1) * B(2,2))
5330      RETURN : REM from determinant subroutine
6999      REM
7000      REM Plot Y and Y2 as a function of X, Apr 15, 81
          (Continue with lines 7010−7980 of Figure 5.4.)

7990      RETURN : REM end of plot
9999      END
```

Figure 7.1: A Parabolic Least-Squares Fit (cont.)

Notice that the plotting subroutine that we used in Chapter 5 is called by the main program. However, the source program is not shown in the listing.

The plotting subroutine requires the array of independent variables to be arranged in increasing or decreasing order. The data generated by the input routine at line 500 are so arranged. However, at a later time, you may want to substitute other data that are not sorted. In this case, you can include one of the sorting subroutines developed in Chapter 6.

Running The Program

Type up this program and run it. The results should look like Figure 7.2. The correlation coefficient is close to unity, indicating that the parabolic equation can produce a good fit to the data. The resulting equation is:

$$y = -7.827 + 10.59x - 1.083x^2$$

```
        Parabolic  least-squares  fit  by  Cramer's  rule
               X        Y      Y Calc
        1     1.0      2.07      1.68
        2     2.0      8.60      9.02
        3     3.0     14.42     14.20
        4     4.0     15.80     17.21
        5     5.0     18.92     18.05
        6     6.0     17.96     16.73
        7     7.0     12.98     13.24
        8     8.0      6.45      7.58
        9     9.0      0.27     -0.24

    Coefficients
        -7.8267          Constant  term
        10.5901
        -1.0830

    Correlation  coefficient  is  .991547


    1               *+
    -
    2                           +*
    -
    3                                       *
    -
    4                                    +     *
    -
    5                                         *   +
    -
    6                                       *   +
    -
    7                               +*
    -
    8                   +     *
    -
    9           *+
              ^        ^         ^         ^         ^         ^
          -0.2      3.6       7.4      11.3      15.1      18.9
```

Figure 7.2: Output from the Parabolic Curve Fit Program

Now let us consider polynomial equations with orders higher than 2, and nonpolynomial equations. We will see that the approach we have been using to solve for the coefficients becomes less practical as the equation becomes more complex. Therefore, we will want to investigate another method.

CURVE FITS FOR OTHER EQUATIONS

If a higher-order polynomial is chosen, there will be additional coefficients, resulting in additional equations to be solved simultaneously. Equations 3, 4, and 5 can be easily extended in this case. For example, to find the coefficients to the cubic equation:

$$y = A + Bx + Cx^2 + Dx^3$$

the residuals are defined as:

$$r = A + Bx + Cx^2 + Dx^3 - y$$

The residuals are first squared and then summed. The derivative is taken with respect to each of the four variables A, B, C, and D. The resulting four equations are solved simultaneously:

$$An + B\Sigma x + C\Sigma x^2 + D\Sigma x^3 = \Sigma y$$
$$A\Sigma x + B\Sigma x^2 + C\Sigma x^3 + D\Sigma x^4 = \Sigma xy$$
$$A\Sigma x^2 + B\Sigma x^3 + C\Sigma x^4 + D\Sigma x^5 = \Sigma x^2 y$$
$$A\Sigma x^3 + B\Sigma x^4 + C\Sigma x^5 + D\Sigma x^6 = \Sigma x^3 y$$

Nonpolynomial expressions are treated similarly. For example, the general three-term equation is:

$$y = A + Bf(x) + Cg(x)$$

where $f(x)$ and $g(x)$ represent any function of x. The residuals, defined as:

$$r = A + Bf(x) + Cg(x) - y$$

are squared and then summed. The derivatives with respect to the three variables give the three equations:

$$An + B\Sigma f(x) + C\Sigma g(x) = \Sigma y$$
$$A\Sigma f(x) + B\Sigma f(x)^2 + C\Sigma f(x)g(x) = \Sigma f(x)y$$
$$A\Sigma g(x) + B\Sigma f(x)g(x) + C\Sigma g(x)^2 = \Sigma g(x)y$$

While the above approach to curve fitting is correct, it is laborious. Major alterations are needed whenever the curve-fitting equation is changed. For example, if the equation were chosen to be:

$$y = A + Bx + \frac{C}{x^2}$$

then there would need to be statements in the program such as:

SUMX5 = SUMX5 + 1 / XI
SUMX6 = SUMX6 + 1 / (XI * XI)

for calculating the needed sums.

A Direct Solution

A better way to determine the coefficients of the curve-fitting equation is to set up the data in a matrix and a vector. The data matrix and data vector are then converted to a set of simultaneous equations that are solved in the usual way. For example, suppose that we want a linear fit to the equation:

$$y = A + Bx$$

for five sets of x-y data. The data vector, in this case, would simply be the vector of y values. The data matrix would look like this:

$$\begin{bmatrix} 1 & x_1 \\ 1 & x_2 \\ 1 & x_3 \\ 1 & x_4 \\ 1 & x_5 \end{bmatrix}$$

Each row of the data matrix corresponds to one of the data points, whereas each column of the matrix corresponds to one term of the equation. Consequently, the data matrix has 5 rows and 2 columns. Column 1 of the data matrix contains only the value of 1, since that is the corresponding function of x (i.e., x^0) in the first term of the equation. Column 2 contains the values of x, because that is the function of x in the second term of the equation.

The data matrix for a parabolic fit would have three columns. The first two columns would be the same as for the straight-line fit. The third column, however, would contain the square of each x value. If, on the other hand, we chose an equation like:

$$\ln p = A + \frac{B}{t} + C \ln t$$

then the first column of the data matrix would contain the value of 1, the second column would have the reciprocal of the data, and the third column would contain the logarithm of the data. The data vector in this case would have the logarithm of p.

The rectangular data matrix is converted to a square matrix by using a simple operation. The transpose of the data matrix is multiplied by the matrix itself to produce the coefficient matrix. For the straight-line fit of five sets of x-y data, the operation is:

$$\begin{bmatrix} 1 & 1 & 1 & 1 & 1 \\ x_1 & x_2 & x_3 & x_4 & x_5 \end{bmatrix} \begin{bmatrix} 1 & x_1 \\ 1 & x_2 \\ 1 & x_3 \\ 1 & x_4 \\ 1 & x_5 \end{bmatrix}$$

The result is a 2-by-2 matrix containing the required sums of x:

$$\begin{bmatrix} n & \Sigma x \\ \Sigma x & \Sigma x^2 \end{bmatrix}$$

The product of the data vector (considered as a row vector) and the data matrix:

$$\begin{vmatrix} y_1 & y_2 & y_3 & y_4 & y_5 \end{vmatrix} \begin{bmatrix} 1 & x_1 \\ 1 & x_2 \\ 1 & x_3 \\ 1 & x_4 \\ 1 & x_5 \end{bmatrix}$$

gives the required constant vector of length 2:

$$\begin{bmatrix} \Sigma y & \Sigma xy \end{bmatrix}$$

We will now examine a BASIC implementation of this approach. We will also want to incorporate into the program a general technique for calculating the standard error. As we predicted in Chapter 4, the Gauss-Jordan method of solving simultaneous equations will prove to be an important tool to use here because it supplies the *inverse* of the coefficient matrix.

BASIC PROGRAM: THE MATRIX APPROACH TO CURVE FITTING

Figure 7.3 gives a curve-fitting program that utilizes this matrix approach. The data matrix and data vector are set up in the input subroutine at line 500. Then the subroutine at line 800 converts the data matrix, X, and data vector, y, into the square coefficient matrix, A, and the constant vector, **g**. The operations are:

$$X^T X = A$$

and

$$yX = \mathbf{g}$$

The data vector is multiplied by the data matrix to produce the constant vector. The solution vector:

$$A^{-1}\mathbf{g} = \mathbf{b}$$

can be obtained by using any of the routines we developed in Chapter 4 for the solution of simultaneous equations.

The linear curve-fitting program developed in Chapter 5 presented the standard errors along with the corresponding elements of the solution to the approximating function. The standard errors were obtained from the standard error of the estimate and the summation of x and x^2. We will use a more general technique at this point.

The standard errors are readily obtained from the inverse of the coefficient matrix, A. The value corresponding to ith term of the approximating function is the product of the standard error of the estimate (SEE) and the square root of the ith term of the major diagonal of the inverse:

$$\begin{bmatrix} a_{11} & & \\ & a_{22} & \\ & & a_{33} \end{bmatrix}^{-1}$$

The BASIC expression is

SIGMA(I%) = SEE * SQR(A(I%,I%))

Of the several methods we previously considered for the simultaneous solution of linear equations, only the Gauss-Jordan method generated the inverse of the coefficient matrix. Since we need this inverse to determine the errors on the elements of the solution vector, the Gauss-Jordan method is the natural choice. Consequently, we will use this method for the remaining curve-fitting programs in this chapter.

The standard errors on the coefficients can be used to determine the confidence intervals for the corresponding coefficients. In addition, the standard errors can alert us to the possibility of ill conditioning. We discussed ill-conditioned matrices in Chapter 4. They are more likely to be encountered during the solution of simultaneous equations than in curve fitting. Nevertheless, we will want to watch for such a problem. The standard errors are derived from the square roots of the diagonal elements of the inverted matrix. Large differences in these elements suggest ill conditioning. Therefore, if the squares of the errors are many orders apart, then ill conditioning may be present. The last program in this chapter demonstrates ill conditioning.

The program shown in Figure 7.3 is similar to the one given in Figure 7.1, but the solution to the parabolic equation is determined by the more general matrix method. Three previously developed routines are utilized. The matrix multiplication subroutine from Chapter 3 is shown in the listing. In addition, the Gauss-Jordan subroutine (developed in Chapter 4) and the plotting routine are utilized.

```
10   REM Parabolic fit by Gauss—Jordan, Apr 19, 1981
11   REM identifiers
13   REM      C1      COEF           solution vector
14   REM      C3      CORREL         correlation coefficient
15   REM      E2      SIGMA          vector of errors
16   REM      E5      SEE            std error of estimate
17   REM      L3%     NLIN%          number of plot lines
18   REM      M1%     MAX%           maximum length
19   REM      N1%     NROW%          number of rows
20   REM      N2%     NCOL%          number of columns
21   REM      R3      RESID          vector of residuals
22   REM      S7      SUMY           sum of y
23   REM      S8      SUMY2          sum y squared
24   REM      T6      SRS            sum residuals squared
25   REM      Y2      YCALC          calculated y
26   REM end of identifiers
30   CLEAR 200
40   A$ = '' ###   ##.#   ###.##   ###.##   ###.##''
50   C$ = ''  ##.####      ##.###^^^''
60   M1% = 35
70   DIM Z(4), A(4,4), C1(4), Y(35), U(35,4)
80   DIM W(4,1), B(4,4), I2%(4,3), X(35), Y1(35)
90   DIM Y2(35), R3(35), E2(4)
100  REM
110  PRINT
120  PRINT '' Parabolic least—squares fit by'';
```

Figure 7.3: A Parabolic Least-Squares Fit Using Gauss-Jordan Elimination ⎯

```
130    PRINT " Gauss—Jordan elimination"
140    GOSUB 500 : REM get the data
150    REM sort the data
160    GOSUB 800 : REM set up the matrix
170    GOSUB 4000 : REM square up the matrix
180    GOSUB 5000 : REM Gauss—Jordan solution
190    GOSUB 1000 : REM Print results
200    GOSUB 7000 : REM plot data
210    GOTO 9999 : REM done
500    REM get the data
510    N1% = 9
520    N2% = 3
530    FOR I% = 1 TO N1%
540        X(I%) = I%
550        READ Y1(I%)
560    NEXT I%
570    L3% = (N1% − 1) * 2 + 1
630    REM Y data
640    DATA 2.07, 8.6, 14.42, 15.8
650    DATA 18.92, 17.96, 12.98, 6.45, 0.27
660    RETURN : REM from input routine
800    REM set up the data matrix
810    FOR I% = 1 TO N1%
820        U(I%,1) = 1
830        FOR J% = 2 TO N2%
840            U(I%,J%) = U(I%,J%−1) * X(I%)
850        NEXT J%
860        Y(I%) = Y1(I%)
870    NEXT I%
880    RETURN : REM from setting up data matrix
1000   REM Calculate residuals and print results
1010   S7 = 0
1020   S8 = 0
1030   T6 = 0
```

Figure 7.3:
A Parabolic Least-Squares Fit Using Gauss-Jordan Elimination (cont.)

```
1040    FOR I% = 1 TO N1%
1050      Y2 = 0
1060      FOR J% = 1 TO N2%
1070        Y2 = Y2 + C1(J%) * U(I%,J%)
1080      NEXT J%
1090      R3(I%) = Y2 - Y(I%)
1100      Y2(I%) = Y2
1110      T6 = T6 + R3(I%) * R3(I%)
1120      S7 = S7 + Y(I%)
1130      S8 = S8 + Y(I%) * Y(I%)
1140    NEXT I%
1150    C3 = SQR(1 - T6 / (S8 - S7 * S7/N1%))
1160    IF (N1% = N2%) THEN E5 = SQR(T6)
1170    IF (N1% <> N2%) THEN E5 = SQR(T6/(N1% - N2%))
1180    FOR J% = 1 TO N2%
1190      E2(J%) = E5 * SQR(ABS(B(J%,J%)))
1200    NEXT J%
1210    PRINT "          X        Y       Y Calc      Resid"
1220    FOR I% = 1 TO N1%
1230      PRINT USING A$; I%; X(I%), Y(I%), Y2(I%), R3(I%)
1240    NEXT I%
1250    PRINT
1260    PRINT "Coefficients        Errors"
1270    PRINT USING C$; C1(1), E2(1);
1280    PRINT "   Constant  term"
1290    FOR I% = 2 TO N2%
1300      PRINT USING C$; C1(I%), E2(I%)
1310    NEXT I%
1320    PRINT
1330    PRINT "Correlation  coefficient  is"; C3
1340    RETURN : REM from printout
4000    REM U and Y converted to A and Z
4011    REM identifiers
4013    REM    N1%    NROW%      number of rows
```

Figure 7.3:
A Parabolic Least-Squares Fit Using Gauss-Jordan Elimination (cont.)

```
4014   REM      N2%    NCOL%      number of columns
4015   REM end of identifiers
4016   REM
4020   FOR K% = 1 TO N2%
4030      FOR L% = 1 TO K%
4040         A(K%,L%) = 0
4050         FOR I% = 1 TO N1%
4060            A(K%,L%) = A(K%,L%) + U(I%,L%) * U(I%,K%)
4070            IF (K% <> L%) THEN A(L%,K%) = A(K%,L%)
4080         NEXT I%
4090      NEXT L%
4100      Z(K%) = 0
4110      FOR I% = 1 TO N1%
4120         Z(K%) = Z(K%) + Y(I%) * U(I%,K%)
4130      NEXT I%
4140   NEXT K%
4150   RETURN : REM from square
5000   REM Gauss — Jordan matrix inversion and solution
       (Continue with lines 5010 – 6130 of Figure 4.6.)
6140   RETURN : REM from Gauss — Jordan subroutine
7000   REM Plot Y and YCALC as a function of X
       (Continue with lines 7010 – 7980 of Figure 5.4.)
7990   RETURN : REM end of plot
9999   END
```

Figure 7.3:
A Parabolic Least-Squares Fit Using Gauss-Jordan Elimination (cont.)

Running the Program

Alter the first program in this chapter or create a new one so that it looks like the one listed in Figure 7.3. Be sure to include the Gauss-Jordan and plotting routines. Run the program and compare the output with Figure 7.4. The results should be the same as for the previous program except that the residuals and errors are given in the new version.

In the next section the matrix approach will permit us to try out different orders of polynomial equations to fit any given set of data. To develop a sense of the full power of this tool, we will run this new program several times on one set of data. By comparing the resulting plotted curves and correlation coefficients, we will find the best polynomial order to fit our data.

```
Parabolic  least-squares  fit  by  Gauss-Jordan  elimination

            X          Y      Y Calc      Resid
    1      1.0       2.07      1.68      -0.39
    2      2.0       8.60      9.02       0.42
    3      3.0      14.42     14.20      -0.22
    4      4.0      15.80     17.21       1.41
    5      5.0      18.92     18.05      -0.87
    6      6.0      17.96     16.73      -1.23
    7      7.0      12.98     13.24       0.26
    8      8.0       6.45      7.58       1.13
    9      9.0       0.27     -0.24      -0.51


Coefficients         Errors
   -7.8266          1.298E+00    Constant  term
   10.5900          5.960E-01
   -1.0830          5.813E-02

Correlation  coefficient  is  .991547
```

Figure 7.4: Output: An Alternate Version of a Parabolic Least-Squares Curve Fit

BASIC PROGRAM: ADJUSTING THE ORDER OF THE POLYNOMIAL

One of the advantages of the new version of our curve-fitting program is that it is easy to change both the number of rows, corresponding to the number of data points, and the number of columns, corresponding to the number of polynomial terms in the curve-fitting equation. Consequently, for this third version, we will proceed one step further in "generalizing" our program. We will input the order of the polynomial equation from the console. The order, of course, is one smaller than the number of terms in the equation. Make a copy of the previous program and alter it so that it looks like the one shown in Figure 7.5.

10	**REM** Parabolic fit by Gauss — Jordan, Apr 21, 1981			
11	**REM** identifiers			
13	**REM**	C1	COEF	solution vector
14	**REM**	C3	CORREL	correlation coefficient
15	**REM**	E2	SIGMA	vector of errors
16	**REM**	E5	SEE	std error of estimate
19	**REM**	L3%	NLIN%	number of plot lines

Figure 7.5: Input Polynomial Order from the Console

```
 20   REM      M1%     MAX%       maximum length
 21   REM      N1%     NROW%      number of rows
 22   REM      N2%     NCOL%      number of columns
 23   REM      N4%     NTRY%      plot entry flag
 24   REM      R3      RESID      vector of residuals
 25   REM      S7      SUMY       sum of y
 26   REM      S8      SUMY2      sum y squared
 27   REM      T6      SRS        sum residuals squared
 28   REM      Y2      YCALC      calculated y
 29   REM end of identifiers
 30   CLEAR 200
 40   A$ = " ###   ##.#   ###.##   ###.##   ###.##"
 50   C$ = "  ##.####      ##.###^^^^"
 60   M1% = 35
 70   DIM Z(4), A(4,4), C1(4), Y(35), U(35,4)
 80   DIM W(4,1), B(4,4), I2%(4,3), X(35), Y1(35)
 90   DIM Y2(35), R3(35), E2(4)
100   REM
110   PRINT
120   PRINT " Parabolic  least—squares  fit"
130   REM
140   GOSUB 500 : REM get the data
150   REM sort the data
160   GOSUB 800 : REM set up the matrix
170   GOSUB 4000 : REM square up the matrix
180   GOSUB 5000 : REM Gauss—Jordan solution
190   GOSUB 1000 : REM Print results
200   GOSUB 7000 : REM plot data
210   GOTO  100 : REM next
500   REM get the data
510   N1% = 9
520   INPUT "Polynomial  order"; N2%
530   IF (N2% > 3) THEN 520
540   IF (N2% < 1) THEN 9999
```

Figure 7.5: Input Polynomial Order from the Console (cont.)

```
550   N2% = N2% + 1
560   L3% = (N1% − 1) * 2 + 1
570   IF (N4% = 1) THEN 620
580   FOR I% = 1 TO N1%
590      X(I%) = I%
600      READ Y1(I%)
610   NEXT I%
620   RETURN : REM from input routine
630   REM Y data
640   DATA  2.07, 8.6, 14.42, 15.8
650   DATA 18.92, 17.96, 12.98, 6.45, 0.27
800   REM set up the data matrix
810   FOR I% = 1 TO N1%
820      U(I%,1) = 1
830      FOR J% = 2 TO N2%
840         U(I%,J%) = U(I%,J%−1) * X(I%)
850      NEXT J%
860      Y(I%) = Y1(I%)
870   NEXT I%
880   RETURN : REM from setting up data matrix
1000  REM Calculate residuals and print results
      (Continue with lines 1010−1330 of Figure 7.3.)
1340  RETURN : REM from printout
4000  REM U and Y converted to A and Z
      (Continue with lines 4010−4140 of Figure 7.3.)
4150  RETURN : REM from square
5000  REM Gauss−Jordan matrix inversion and solution
      (Continue with lines 5010−6130 of Figure 4.6.)
6140  RETURN : REM from Gauss−Jordan subroutine
7000  REM Plot Y and YCALC as a function of X
      (Continue with lines 7010−7980 of Figure 5.4.)
7990  RETURN : REM end of plot
9999  END
```

Figure 7.5: Input Polynomial Order from the Console (cont.)

Comparing Runs of the Program

Run the program and input a value of 2 for the polynomial order. The result should again be the same as for the two previous versions. This time, however, the program will cycle and ask for the polynomial order again. Give a value of 1 the second time. The results (shown in Figure 7.6) are the best straight line through the curved set of data.

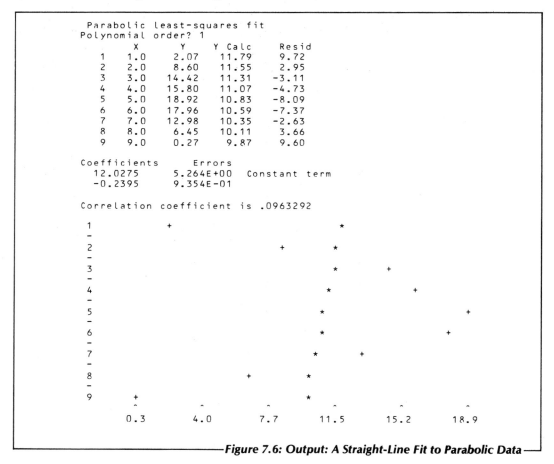

Figure 7.6: Output: A Straight-Line Fit to Parabolic Data

The two coefficients represent the equation:

$$y = 12.028 - 0.240x$$

Notice that about half of the data points are on one side of the straight line and half are on the other. The straight line goes through the points, as best it can. The correlation coefficient, however, is less than 0.1. This relatively small value indicates that the straight-line fit is not very good.

Give an order of 3 for the third cycle. This will produce a fit corresponding to the cubic equation:

$$y = -7.2 + 10.0x - 0.946x^2 - 0.00915x^3$$

From the value of the correlation coefficient, it can be seen that the resulting curve describes the data no better than the parabolic fit. One should always use the lowest-order equation that reasonably fits the data. Therefore, the parabola would be the ideal choice in this case.

With this present version, we can choose a polynomial with order up to 3. This restriction is necessary because we have set the maximum number of matrix columns to be 4 in the dimension statements at the beginning of the program. If a higher order polynomial is necessary, change the value of 4 in each dimension statement to the correspondingly higher value.

The program is terminated by entering a polynomial order that is zero or negative. You may want to alter the program so that it asks for the number of terms, rather than the polynomial order. In that case, line 560:

    560 N2% = N2% + 1

should be removed.

In the next three sections we will look at actual experimental applications of curve fitting involving the heat capacity of oxygen, the vapor pressure of liquid lead, and the properties of superheated steam.

BASIC PROGRAM: THE HEAT-CAPACITY EQUATION

Heat capacity is a measure of how much the temperature of a body will increase when a given amount of heat is added. Experimentally determined data are commonly fitted to the equation:

$$C_p = A + BT + \frac{C}{T^2}$$

where C_p is the heat capacity in units of energy per degree, and T is the absolute temperature. The coefficients are again A, B, and C.

Make a copy of the curve-fitting program given in Figure 7.3 (rather than Figure 7.5, which is designed for adjusting the order). Alter the input routine at lines 500-660 so it looks like the version shown in Figure 7.7. As before, the subroutine starting at line 800 fills the first column of the data matrix with the value of unity, and the second column with the temperature (the independent variable). But now the third column contains the reciprocal of the temperature squared.

```
 10   REM Curve fit to heat capacity, Apr 19, 1981
 11   REM identifiers
 13   REM      C1      COEF          solution vector
 14   REM      C3      CORREL        correlation coefficient
 15   REM      E2      SIGMA         vector of errors
 16   REM      E5      SEE           std error of estimate
 17   REM      L3%     NLIN%         number of plot lines
 18   REM      M1%     MAX%          maximum length
 19   REM      N1%     NROW%         number of rows
 20   REM      N2%     NCOL%         number of columns
 21   REM      R3      RESID         vector of residuals
 22   REM      S7      SUMY          sum of y
 23   REM      S8      SUMY2         sum y squared
 24   REM      T6      SRS           sum residuals squared
 25   REM      Y2      YCALC         calculated y
 26   REM end of identifiers
 30   CLEAR 200
 40   A$ = '' ###   ####   ###.##   ###.##   ###.##''
 50   C$ = ''##.####^^^^      ##.###^^^^''
 60   M1% = 35
 70   DIM Z(4), A(4,4), C1(4), Y(35), U(35,4)
 80   DIM W(4,1), B(4,4), I2%(4,3), X(35), Y1(35)
 90   DIM Y2(35), R3(35), E2(4)
100   REM
110   PRINT
120   PRINT '' Least—squares  fit  to  heat  capacity''
130   REM
140      GOSUB 500 : REM get the data
150      REM sort the data
160      GOSUB 800 : REM set up the matrix
170      GOSUB 4000 : REM square up the matrix
180      GOSUB 5000 : REM Gauss—Jordan solution
190      GOSUB 1000 : REM Print results
200      GOSUB 7000 : REM plot data
210      GOTO 9999 : REM done
```

Figure 7.7: Least-Squares Fit to the Heat-Capacity Equation

```
500    REM get the data
510    N1% = 10
520    N2% = 3
530    FOR I% = 1 TO N1%
540        X(I%) = (I% + 2) * 100
550        READ Y1(I%)
560    NEXT I%
570    L3% = (N1% - 1) * 2 + 1
630    REM Y data
640    DATA 7.02, 7.2, 7.43, 7.67, 7.88
650    DATA 8.06, 8.21, 8.34, 8.44, 8.53
660    RETURN : REM from input routine
800    REM set up data matrix for heat capacity
810    FOR I% = 1 TO N1%
820        U(I%,1) = 1
830        U(I%,2) = X(I%)
840        U(I%,3) = 1 / (X(I%) * X(I%))
850    Y(I%) = Y1(I%)
860    NEXT I%
870    REM
880    RETURN : REM from setting up data matrix
1000   REM Calculate residuals and print results
       (Continue with lines 1010 - 1330 of Figure 7.3.)

1340   RETURN : REM from printout
4000   REM U and Y converted to A and Z
       (Continue with lines 4010 - 4140 of Figure 7.3.)

4150   RETURN : REM from square
5000   REM Gauss - Jordan matrix inversion and solution
       (Continue with lines 5010 - 6130 of Figure 4.6.)

6140   RETURN : REM from Gauss - Jordan subroutine
7000   REM Plot Y and YCALC as a function of X
       (Continue with lines 7010 - 7980 of Figure 5.4.)

7990   RETURN : REM from plot
9999   END
```

Figure 7.7: Least-Squares Fit to the Heat-Capacity Equation (cont.)

The data represent the heat capacity of oxygen over the temperature range of 300 to 1200 degrees Kelvin. Compile the program and run it. The output should look like Figure 7.8. The resulting equation is:

$$C_p = 6.9 + 0.00143T - \frac{32610}{T^2}$$

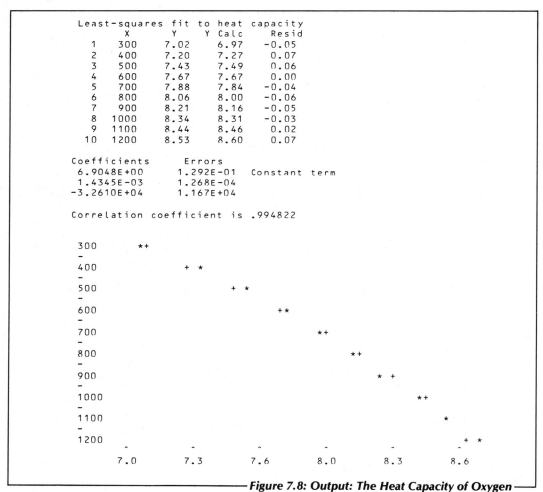

```
        Least-squares  fit  to  heat  capacity
              X         Y      Y Calc     Resid
        1    300      7.02     6.97      -0.05
        2    400      7.20     7.27       0.07
        3    500      7.43     7.49       0.06
        4    600      7.67     7.67       0.00
        5    700      7.88     7.84      -0.04
        6    800      8.06     8.00      -0.06
        7    900      8.21     8.16      -0.05
        8   1000      8.34     8.31      -0.03
        9   1100      8.44     8.46       0.02
       10   1200      8.53     8.60       0.07

   Coefficients        Errors
   6.9048E+00         1.292E-01    Constant  term
   1.4345E-03         1.268E-04
  -3.2610E+04         1.167E+04

   Correlation  coefficient  is  .994822


       300        *+
        -
       400               +  *
        -
       500                    +  *
        -
       600                        +*
        -
       700                             *+
        -
       800                               *+
        -
       900                                 *  +
        -
      1000                                   *+
        -
      1100                                     *
        -
      1200                                       +  *
               ^       ^        ^        ^        ^       ^
              7.0     7.3      7.6      8.0      8.3     8.6
```

Figure 7.8: Output: The Heat Capacity of Oxygen

BASIC PROGRAM: THE VAPOR PRESSURE EQUATION

When a gas or vapor is in equilibrium with its own liquid or solid, we say that it is saturated. For pure materials, the saturation pressure is a single-valued function of the temperature. A commonly used equation to express the relationship between the saturation pressure and the saturation temperature is:

$$\log P = A + \frac{B}{T} + C \log T$$

In this equation, P is the pressure and T is the absolute temperature. The coefficients are A, B, and C. Either the natural logarithm or the common logarithm is used.

Make a copy of the program shown in Figure 7.7. Alter the subroutines starting at lines 500 and 800 so they look like the versions shown in Figure 7.9. The subroutine at line 800 sets the first column of the matrix to unity as usual. Column 2 is then filled with the reciprocal of the independent variable, the temperature. Column 3 gets the logarithm of the temperature. Run the new program. The results should look like Figure 7.10. The data represent the vapor pressure of liquid lead over the temperature range of 700 to 1600 degrees Kelvin. The corresponding equation is:

$$\ln P = 18 - \frac{23200}{T} - .89 \ln T$$

when the pressure is in atmospheres and the natural logarithm is used. The correlation coefficient of 1 indicates a very good fit.

10	**REM** Curve fit to lead vapor pressure, Apr 19, 1981		
11	**REM** identifiers		
13	**REM**	C1	COEF	solution vector
14	**REM**	C3	CORREL	correlation coefficient
15	**REM**	E2	SIGMA	vector of errors
16	**REM**	E5	SEE	std error of estimate
19	**REM**	L3%	NLIN%	number of plot lines
20	**REM**	M1%	MAX%	maximum length
21	**REM**	N1%	NROW%	number of rows
22	**REM**	N2%	NCOL%	number of columns
23	**REM**	R3	RESID	vector of residuals
24	**REM**	S7	SUMY	sum of y
25	**REM**	S8	SUMY2	sum y squared
26	**REM**	T6	SRS	sum residuals squared
27	**REM**	Y2	YCALC	calculated y
28	**REM** end of identifiers		
30	**CLEAR** 200		

Figure 7.9: Least-Squares Fit to the Vapor Pressure Equation

```
 40    A$ = '' ###   ####   ###.##   ###.##   ###.##''
 50    C$ = ''##.####^^^^      ##.###^^^^''
 60    M1% = 35
 70    DIM Z(4), A(4,4), C1(4), Y(35), U(35,4)
 80    DIM W(4,1), B(4,4), I2%(4,3), X(35), Y1(35)
 90    DIM Y2(35), R3(35), E2(4)
100    REM
110    PRINT
120    PRINT '' Least—squares fit for lead vapor pressure''
130    REM
140    GOSUB 500 : REM get the data
150    REM sort the data
160    GOSUB 800 : REM set up the matrix
170    GOSUB 4000 : REM square up the matrix
180    GOSUB 5000 : REM Gauss—Jordan solution
190    GOSUB 1000 : REM Print results
200    GOSUB 7000 : REM plot data
210    GOTO 9999 : REM done
500    REM get the data
510    N1% = 10
520    N2% = 3
530    FOR I% = 1 TO N1%
540       X(I%) = (I% + 6) * 100
550       READ Y1(I%)
555       Y1(I%) = LOG(Y1(I%))
560    NEXT I%
570    L3% = (N1% − 1) * 2 + 1
630    REM Y data
640    DATA  1.0E−9, 5.598E−8
642    DATA  1.234E−6, 1.507E−5
646    DATA  1.138E−4, 6.067E−4
648    DATA  2.512E−3, 8.337E−3
650    DATA  2.371E−2, 5.875E−2
660    RETURN : REM from input routine
800    REM set up data matrix for vapor pressure
```

Figure 7.9: Least-Squares Fit to the Vapor Pressure Equation (cont.)

```
810    FOR I% = 1 TO N1%
820        U(I%,1) = 1
830        U(I%,2) = 1 / X(I%)
840        U(I%,3) = LOG(X(I%))
850        Y(I%) = Y1(I%)
860    NEXT I%
870    REM
880    RETURN : REM from setting up data matrix
1000   REM Calculate residuals and print results
           (Continue with lines 1010–1330 of Figure 7.3.)
1340   RETURN : REM from printout
4000   REM U and Y converted to A and Z
           (Continue with lines 4010–4140 of Figure 7.3.)
4150   RETURN : REM from square
5000   REM Gauss–Jordan matrix inversion and solution
           (Continue with lines 5010–6130 of Figure 4.6.)
6140   RETURN : REM from Gauss–Jordan subroutine
7000   REM Plot Y and YCALC as a function of X
           (Continue with lines 7010–7980 of Figure 5.4.)
7990   RETURN : REM from plot
9999   END
```

Figure 7.9: Least-Squares Fit to the Vapor Pressure Equation (cont.)

```
      Least-squares fit for lead vapor pressure
             X       Y     Y Calc     Resid
        1    700  -20.72   -20.72    -0.00
        2    800  -16.70   -16.70    -0.01
        3    900  -13.61   -13.59     0.02
        4   1000  -11.10   -11.11    -0.00
        5   1100   -9.08    -9.09    -0.00
        6   1200   -7.41    -7.41     0.00
        7   1300   -5.99    -5.99    -0.01
        8   1400   -4.79    -4.78     0.00
        9   1500   -3.74    -3.74    -0.00
       10   1600   -2.83    -2.83     0.00

      Coefficients         Errors
        1.8251E+01        5.946E-01   Constant term
       -2.3184E+04        7.748E+01
       -8.9386E-01        7.455E-02

      Correlation coefficient is 1
```

Figure 7.10: Output: The Vapor Pressure of Lead

Notice that the original pressures in the input subroutine are given in atmospheres. These values are then converted to the logarithm of the pressure. The x and y values that are printed are actually the temperature and the logarithm of the pressure. It is left as an exercise for the reader to include the original pressure values in the final printout.

In the final example we will encounter an equation that breaks the restriction we established at the beginning of the chapter. One of the unknown coefficients is in a nonlinear form—specifically, an exponential. Later, in Chapter 10, we will study algorithms for handling such an equation, but here we will have to be satisfied with a less elegant solution. We will estimate values for the exponent until we find the optimum correlation coefficient.

A THREE-VARIABLE EQUATION

In the previous section, we considered an equation of state for a saturated gas. The pressure is a function of temperature for this condition. We will now consider an equation of state for a superheated gas. In this case, the temperature is above the saturation temperature or, put another way, the pressure is below the saturation pressure. Since temperature and pressure are independent variables under this condition, we can express the volume as a function of both the temperature and the pressure.

For an ideal gas, the equation of state is:

$$PV = RT$$

where P is the pressure, V is the molar volume, R is the gas constant, and T is the temperature. But as the pressure increases, the behavior of the gas becomes less and less ideal. A common non-ideal equation of state is:

$$PV = A + BP + CP^2 + DP^3 + \ldots$$

It can be seen that this equation is just a power series expansion in pressure.

Be careful not to confuse the gas constant (R) with the vector of residuals (r). In the BASIC program, the symbol G2 is used for the gas constant and the symbol R3 is used for the vector of residuals.

Since all gases become ideal as the pressure is reduced, the equation of state should merge smoothly with the ideal gas equation as the pressure approaches zero. For this reason, the value of A in the above equation must be equal to RT. Because fewer polynomial terms are needed at lower values of pressure, the equation of state can often be written as:

$$PV = RT + BP + CP^2$$

The determination of the coefficients B and C is straightforward when the temperature is constant. However, if temperature is also a variable, then the coefficients B and C need to be functions of temperature.

Several different equations of state are in common use. All of them, however, are empirical. As an example of a three-variable equation, consider the expression:

$$PV = RT + \frac{BP}{T^n} + CP^2$$

In this case, coefficient C is not a function of T, but the original coefficient B has become the function:

$$\frac{B}{T^n}$$

This equation has a nonlinear coefficient, n, so we cannot obtain a solution by the methods discussed in this chapter. However, if an estimate is made for the coefficient n, the remaining linear coefficients can be determined. We will begin with an estimate of unity for coefficient n and determine the other coefficients by our usual least-squares method. We can then observe how well the resulting equation represents the original data. Then we will change the value of n and see whether the new equation becomes better or worse.

BASIC PROGRAM: AN EQUATION OF STATE FOR STEAM

The program given in Figure 7.11 can be used to find the coefficients B and C for the published properties of steam. Since the coefficient of the first term on the right is unity, the equation has been rearranged to give:

$$PV - RT = \frac{BP}{T^n} + CP^2$$

The data are defined in the input routine starting at line 500 as usual. Temperature is given in degrees Fahrenheit, the pressure in pounds per square inch, and the specific volume in cubic feet per pound mass. The temperature data are converted to the absolute Rankine scale by the addition of 460. Pressures are left in pounds per square inch. But then the gas constant, R, which has a value of 85.76 for steam, is divided by 144 square inches per square foot.

The data matrix is set up starting at line 800, as usual. The first column of the matrix contains the pressure divided by the nth power of the temperature. Column 2 contains the square of the pressure. The y vector has the value $PV - RT$.

Notice that this is the first time that we did *not* put the value of unity in the first column of the matrix. We could, of course, divide the equation by the pressure. The right-hand side would then look like a first-order, straight-line fit. But this might cause trouble when pressure became very small.

Comparing Runs of the Program to Determine the Value of *n*

Type up the program given in Figure 7.11 and execute it. Notice that the exponent *n* is initially chosen to be unity. The results should look like Figure 7.12.

```
10   REM Three—variable curve—fit to steam, Apr 21, 81
11   REM identifiers
13   REM      C1      COEF          solution vector
14   REM      C3      CORREL        correlation coefficient
15   REM      E2      SIGMA         vector of errors
16   REM      E5      SEE           std error of estimate
17   REM      G2      GAS           gas constant
18   REM      M1%     MAX%          maximum length
19   REM      N1%     NROW%         number of rows
20   REM      N2%     NCOL%         number of columns
21   REM      P4      POWER
22   REM      R3      RESID         vector of residuals
23   REM      S7      SUMY          sum of y
24   REM      S8      SUMY2         sum y squared
25   REM      T6      SRS           sum residuals squared
26   REM      Y2      YCALC         calculated y
27   REM end of identifiers
30   REM
40   A$ = " ###   ####    ###   ##.###   ###.##   ###.##   ###.##"
50   C$ = "  ##.####^^^^     ##.###^^^^"
60   M1% = 35
65   G2 = 85.76 : REM gas constant for steam
70   DIM Z(4), A(4,4), C1(4), Y(35), U(35,4)
80   DIM W(4,1), B(4,4), I2%(4,3), X(35), Y1(35)
90   DIM Y2(35), R3(35), E2(4)
```

Figure 7.11: An Equation of State for Steam

```
  95    DIM T(35), P(35), V(35)
 100    REM
 110    PRINT
 120    PRINT " An equation of state for steam"
 130    REM
 140       GOSUB 500 : REM get the data
 150       REM
 160       GOSUB 800 : REM set up the matrix
 170       GOSUB 4000 : REM square up the matrix
 180       GOSUB 5000 : REM Gauss—Jordan solution
 190       GOSUB 1000 : REM Print results
 210       GOTO 9999 : REM done
 220       REM
 500    REM get the data
 510    N1% = 12
 520    N2% = 2
 540    FOR I% = 1 TO N1%
 550       READ T(I%), P(I%), V(I%)
 555       T(I%) = T(I%) + 460 : REM convert to Rankine
 560    NEXT I%
 570    REM
 600    REM data for steam
 605    DATA 400, 120, 4.079
 610    DATA 450, 120, 4.36
 615    DATA 500, 120, 4.633
 620    DATA 400, 140, 3.466
 625    DATA 450, 140, 3.713
 630    DATA 500, 140, 3.952
 635    DATA 400, 160, 3.007
 640    DATA 450, 160, 3.228
 645    DATA 500, 160, 3.44
 650    DATA 400, 180, 2.648
 655    DATA 450, 180, 2.85
 660    DATA 500, 180, 3.042
 670    RETURN : REM from input routine
```

Figure 7.11: An Equation of State for Steam (cont.)

```
800    REM set up the data matrix for heat capacity
810    FOR I% = 1 TO N1%
820        P4 = T(I%)
830        U(I%,1) = P(I%) / P4
840        U(I%,2) = SQR(P(I%))
860        Y(I%) = V(I%) * P(I%) − G2 * T(I%) / 144
870    NEXT I%
880    RETURN : REM from setting up data matrix
1000   REM Calculate residuals and print results
1010   S7 = 0
1020   S8 = 0
1030   T6 = 0
1040   FOR I% = 1 TO N1%
1050       Y2 = 0
1060       FOR J% = 1 TO N2%
1070           Y2 = Y2 + C1(J%) * U(I%,J%)
1080       NEXT J%
1090       R3(I%) = Y2 − Y(I%)
1100       Y2(I%) = Y2
1110       T6 = T6 + R3(I%) * R3(I%)
1120       S7 = S7 + Y(I%)
1130       S8 = S8 + Y(I%) * Y(I%)
1135       R3(I%) = 100 * R3(I%)/Y(I%) : REM convert to percent
1140   NEXT I%
1150   C3 = SQR(1 − T6/(S8 − S7 * S7/N1%))
1160   IF (N1% = N2%) THEN E5 = SQR(T6)
1170   IF (N1% > N2%) THEN E5 = SQR(T6 / (N1% − N2%))
1180   FOR J% = 1 TO N2%
1190       E2(J%) = E5 * SQR(ABS(B(J%,J%)))
1200   NEXT J%
1210   PRINT "          P     T     V      Y     Y CALC    %RES"
1220   FOR I% = 1 TO N1%
1230       PRINT USING A$; I%; P(I%), T(I%), V(I%), Y(I%), Y2(I%), R3(I%)
1240   NEXT I%
1250   PRINT
```

Figure 7.11: An Equation of State for Steam (cont.)

```
1260   PRINT "  Coefficients        Errors"
1270   PRINT USING C$; C1(1), E2(1);
1280   PRINT "  Constant term"
1290   FOR I% = 2 TO N2%
1300      PRINT USING C$; C1(I%), E2(I%)
1310   NEXT I%
1320   PRINT
1330   PRINT "Correlation coefficient is"; C3
1340   RETURN : REM from printout
4000   REM U and Y converted to A and Z
       (Continue with lines 4010−4140 of Figure 7.3.)
4150   RETURN : REM from square
5000   REM Gauss−Jordan matrix inversion and solution
       (Continue with lines 5010−6130 of Figure 4.6.)
6140   RETURN : REM from Gauss−Jordan subroutine
9999   END
```

Figure 7.11: An Equation of State for Steam (cont.)

```
An equation of state for steam
           P       T       V        Y       Y CALC      %RES
    1     120     860    4.079   -22.70    -19.44     -14.36
    2     120     910    4.360   -18.76    -17.43      -7.07
    3     120     960    4.633   -15.77    -15.63      -0.92
    4     140     860    3.466   -26.94    -24.16     -10.30
    5     140     910    3.713   -22.14    -21.82      -1.44
    6     140     960    3.952   -18.45    -19.72       6.85
    7     160     860    3.007   -31.06    -28.98      -6.69
    8     160     910    3.228   -25.48    -26.30       3.24
    9     160     960    3.440   -21.33    -23.90      12.03
   10     180     860    2.648   -35.54    -33.88      -4.67
   11     180     910    2.850   -28.96    -30.86       6.59
   12     180     960    3.042   -24.17    -28.16      16.50

   Coefficients        Errors
   -2.6219E+02         4.759E+01   Constant term
    1.5651E+00         6.495E-01

Correlation coefficient is .918393
```

Figure 7.12:
Output: The Properties of Superheated Steam (The exponent n = 1)

It can be seen from Figure 7.12 that the results are not very good. The correlation coefficient is 92%, but some of the calculated values are more than 10% from the original data. We should therefore try something else. Alter the variable P4 at line 820 so that the exponent n will be 2. That is, make P4 equal to the square of the temperature.

820 P4 = T(I%) * T(I%)

Rerun the program. The output should look like Figure 7.13.

```
       An equation of state for steam
               P       T       V        Y       Y CALC     %RES
        1     120     860    4.079   -22.70    -21.49     -5.32
        2     120     910    4.360   -18.76    -18.03     -3.85
        3     120     960    4.633   -15.77    -15.10     -4.25
        4     140     860    3.466   -26.94    -26.01     -3.44
        5     140     910    3.713   -22.14    -21.98     -0.71
        6     140     960    3.952   -18.45    -18.56      0.58
        7     160     860    3.007   -31.06    -30.59     -1.49
        8     160     910    3.228   -25.48    -25.98      2.00
        9     160     960    3.440   -21.33    -22.08      3.48
       10     180     860    2.648   -35.54    -35.22     -0.88
       11     180     910    2.850   -28.96    -30.04      3.74
       12     180     960    3.042   -24.17    -25.64      6.08

       Coefficients        Errors
       -1.9936E+05        1.194E+04    Constant term
        9.9091E-01        1.803E-01

       Correlation coefficient is .98901
```

Figure 7.13:
Output: The Properties of Superheated Steam (The exponent n = 2)

The resulting curve fit looks better this time. The correlation coefficient is 99%, and the calculated points are within 6% of the data. We will now continue in this way by increasing the exponent to 3. Change P4 to read:

820 P4 = T(I%) * T(I%) * T(I%)

Run the program a third time and compare the output to Figure 7.14.

The result is definitely better. The fitted values are within one percent of the original points and the correlation coefficient is 99.96 percent. It looks as though we should accept these results. But just to be sure, either change the exponent to 4 with the statement:

820 P4 = T(I%) * T(I%) * T(I%) * T(I%)

or define the variable H1:

815 H1 = T(I%) * T(I%)
820 P4 = H1 * H1

```
An equation of state for steam
         P       T       V       Y       Y CALC      %RES
    1   120     860    4.079   -22.70    -22.88       0.78
    2   120     910    4.360   -18.76    -18.80       0.22
    3   120     960    4.633   -15.77    -15.52      -1.59
    4   140     860    3.466   -26.94    -26.97       0.13
    5   140     910    3.713   -22.14    -22.21       0.35
    6   140     960    3.952   -18.45    -18.39      -0.32
    7   160     860    3.007   -31.06    -31.09       0.10
    8   160     910    3.228   -25.48    -25.65       0.68
    9   160     960    3.440   -21.33    -21.28      -0.23
   10   180     860    2.648   -35.54    -35.22      -0.90
   11   180     910    2.850   -28.96    -29.10       0.49
   12   180     960    3.042   -24.17    -24.19       0.06


Coefficients            Errors
  -1.3866E+08           1.502E+06    Constant term
   2.9989E-01           2.520E-02

Correlation coefficient is .99963
```

Figure 7.14:
—*Output: The Properties of Superheated Steam (The exponent n = 3)*—

Run the program again and compare the results to Figure 7.15. It can be seen that we have definitely gone too far. The calculated points are not as close as they were for the previous fit and the correlation coefficient is farther from unity. In fact, the Gauss-Jordan procedure may report that the matrix is singular. The problem is ill conditioning. This can be seen by comparing the squares of the two standard errors. This ratio is greater than twenty orders of magnitude.

```
An equation of state for steam
         P       T       V       Y       Y CALC      %RES
    1   120     860    4.079   -22.70    -23.69       4.38
    2   120     910    4.360   -18.76    -19.32       3.02
    3   120     960    4.633   -15.77    -16.00       1.45
    4   140     860    3.466   -26.94    -27.46       1.94
    5   140     910    3.713   -22.14    -22.36       1.02
    6   140     960    3.952   -18.45    -18.49       0.19
    7   160     860    3.007   -31.06    -31.22       0.51
    8   160     910    3.228   -25.48    -25.39      -0.34
    9   160     960    3.440   -21.33    -20.96      -1.74
   10   180     860    2.648   -35.54    -34.96      -1.62
   11   180     910    2.850   -28.96    -28.41      -1.89
   12   180     960    3.042   -24.17    -23.43      -3.08


Coefficients            Errors
  -9.8474E+10           3.648E+09    Constant term
  -1.9071E-01           6.828E-02

Correlation coefficient is .995714
```

Figure 7.15:
—*Output: The Properties of Superheated Steam (The exponent n = 4)*—

At this point, we could try to improve the fit by choosing noninteger exponents close to the value of 3. The results would show, however, that 3 is best. Another possibility is to make coefficient C a function of temperature. But since the result with an exponent of 3 is reasonably good, we should perhaps be satisfied.

SUMMARY

This chapter has illustrated the programming concept of *generalization*. We have now progressed through several versions of our curve-fitting program:

- a straight-line fit

- a parabolic fit

- a more direct parabolic fit, using the matrix approach

- a general polynomial curve fit in which the polynomial order can be adjusted

- curve fits for nonpolynomial equations.

The restriction on all of these versions is that the coefficients must be linear. Although we found a way around this restriction in the last example of the chapter, we must still develop methods of dealing with nonlinear coefficients.

EXERCISES

7-1: *It is possible to calculate the activity of one component of a binary solution from a knowledge of the activity of the other component. The common logarithm of the activity coefficient, G, (the ratio of activity to mole fraction) is fitted to a power series in mole fraction of the other component. Notice that the series begins with the second-order term.*

$$\text{Log } G_1 = A x_2^2 + B x_2^3 + C x_2^4 + \ldots$$

The activity, A_{Ni}, of nickel in iron-nickel solutions at 1500° Kelvin as a function of the mole fraction of nickel, X_{Ni}, has been reported as:

X_{Ni}	1	.9	.8	.7	.6	.5	.4	.3	.2	.1
A	1	.89	.766	.62	.485	.374	.283	.207	.136	.067

Make a copy of the program given in Figure 7.3. Increase the number of data points, N1%, to 10. Change line 540 so that X is defined as the mole fraction of I%:

540 X(I%) = 0.1 ∗ (I% − 1)

*Change the **DATA** statements at lines 640 and 650 so that Y1 will contain the corresponding nickel activities. Change line 820 so that the power series begins with the second-order term:*

820 U(I%,1) = X(I%) ∗ X(I%)

The dependent variable is defined at line 860 as:

860 Y(I%) = LOG(Y1(I%) / (1.0 − X(I%)))

In this step, the activity coefficient is obtained from a ratio of the activity to the mole fraction. The dependent variable is the logarithm of the ratio. Determine the coefficients A, B, and C.

Answer: $A = -2.128$, $B = 2.319$, $C = -0.548$

7-2: *The vapor pressure of mercury is reported to be:*

Temperature, C	Pressure, torr
0	0.000185
50	0.01267
100	0.2729
150	2.2807
200	17.287
250	74.375
300	246.80
350	672.69
400	1574.1

Perform a least-squares fit on the data to find the coefficients A, B, and C to the equation:

$$\ln P = A + B/T + C \ln T$$

The x data can be generated in the input loop:

540 X(I%) = I% * 50 + 273.15

and the y data can be read from a data statement and converted at that point:

580 **READ** Y1(I%)
590 Y1(I%) = LOG(Y1(I%)) . . .
640 **DATA** 1.85E−4, 1.267E−2 . . .

Use the vapor pressure program developed in Figure 7.9.

Answer: $A = 63.85, B = -12125, C = -6.04$

7-3: *Find the coefficients A, B, and C to:*

$$C_p = A + BT + C/T^2$$

for the heat capacity of graphite. Use the program given in Figure 7.7. The temperatures are given in degrees Kelvin and the heat capacities are given in cal/deg mole.

T	C_p
300	2.08
400	2.85
500	3.50
600	4.03
700	4.43
800	4.75
900	4.98
1000	5.14
1100	5.27
1200	5.42

Answer: $A = 3.37, B = 0.00198, C = -177100$

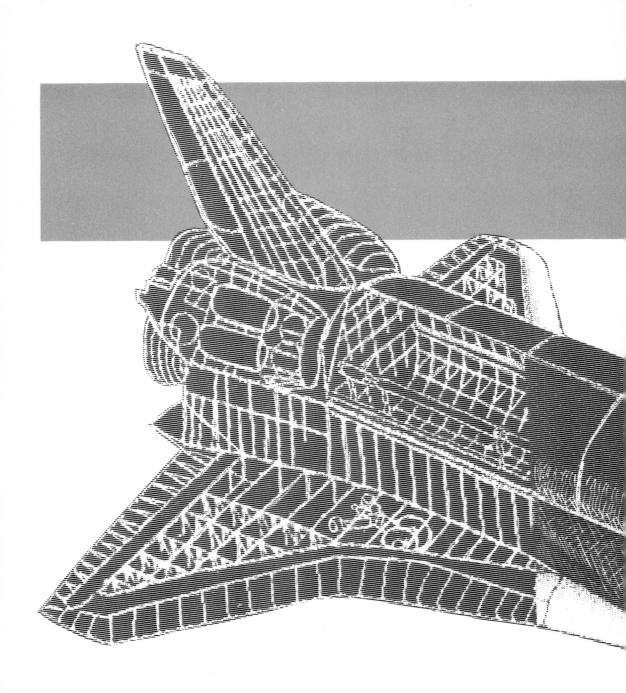

CHAPTER **8**

Solution of Equations by Newton's Method

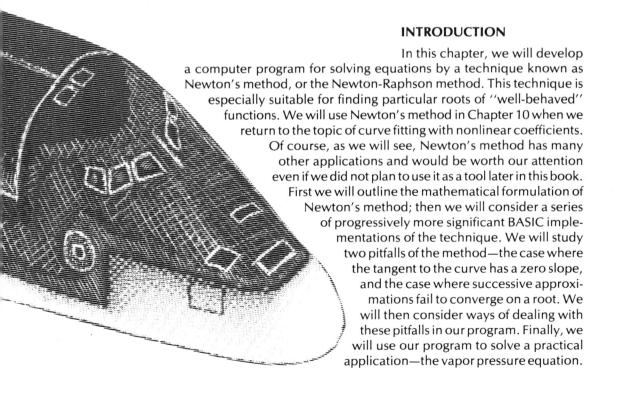

INTRODUCTION

In this chapter, we will develop a computer program for solving equations by a technique known as Newton's method, or the Newton-Raphson method. This technique is especially suitable for finding particular roots of "well-behaved" functions. We will use Newton's method in Chapter 10 when we return to the topic of curve fitting with nonlinear coefficients. Of course, as we will see, Newton's method has many other applications and would be worth our attention even if we did not plan to use it as a tool later in this book. First we will outline the mathematical formulation of Newton's method; then we will consider a series of progressively more significant BASIC implementations of the technique. We will study two pitfalls of the method—the case where the tangent to the curve has a zero slope, and the case where successive approximations fail to converge on a root. We will then consider ways of dealing with these pitfalls in our program. Finally, we will use our program to solve a practical application—the vapor pressure equation.

FORMULATING NEWTON'S METHOD

Let us begin by considering a general equation of the form:

$$f(x) = 0 \tag{1}$$

This equation might have one solution, several solutions, or none at all. That is, there may be particular values of x that make the equation equal to zero. These values are called the *roots* or *solutions* of the equation. For other values of x, the function will not be zero.

Sometimes we can solve such an equation explicitly. For example, the expression:

$$x^2 - 4 = 0$$

can be converted to:

$$x^2 = 4$$

which has the solutions:

$$x = 2 \quad \text{and} \quad x = -2$$

Sometimes, however, an equation cannot be solved so easily. As an example, consider the expression:

$$\ln P = A + \frac{B}{T} + C \ln T$$

This formula, which we will implement into our program at the end of this chapter, can be used to describe the vapor pressure of a material. In this equation P is the pressure, T is the temperature, and A, B, and C are constants that are unique for each substance. For the element lead, the experimentally determined coefficients are

$$A = 18.19$$
$$B = -23180$$
$$C = -0.8858$$

when the pressure is given in atmospheres and the temperature is given in degrees Kelvin.

We can easily find the vapor pressure of lead at, say, 1000° by solving the expression:

$$\ln P = A + \frac{B}{1000} + C \ln 1000$$

But suppose that we want to find the temperature that corresponds to a lead vapor pressure of 0.1 atmospheres. We want to solve the equation:

$$\ln 0.1 = 18.19 - \frac{23180}{T} - 0.8858 \ln T$$

This nonlinear equation cannot be solved explicitly for the temperature. However, we can use an approximation method to calculate the answer to as high a precision as we desire.

For the general case, we write the equation:

$$y = f(x)$$

We are thus interested in the values of x when y is equal to zero; that is, we want to determine the points where the curve of the function crosses the x-axis.

Consider, for example, the curve of the equation:

$$y = f(x) = x^2 - 4$$

This curve crosses the x-axis at two places, $+2$ and -2, as shown in Figure 8.1.

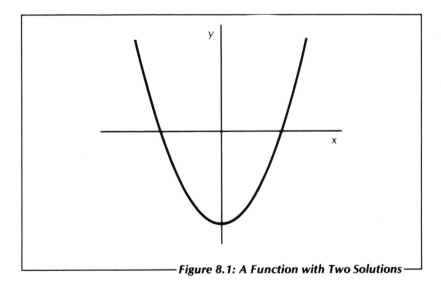

Figure 8.1: A Function with Two Solutions

As a second example, consider the curve of:

$$y = f(x) = x^2$$

This curve is tangent to the x-axis at the origin, corresponding to a single root, $x = 0$, as shown in Figure 8.2.

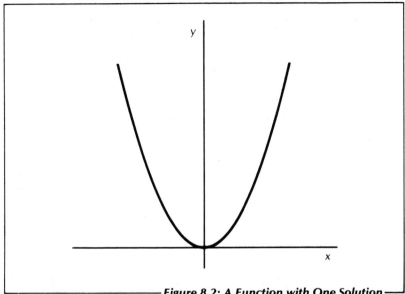

Figure 8.2: A Function with One Solution

Finally, consider the equation:

$$y = f(x) = x^2 + 4$$

shown in Figure 8.3. This equation does not cross the x-axis at all, and so it has no real roots.

Let us explore the behavior of a general function,

$$y = f(x)$$

in the vicinity of a root. We might find that it looks like the curve of Figure 8.4. The function crosses the x-axis at a root because the relationship:

$$y = f(x) = 0$$

is satisfied there.

A Series of tangents to the Curve $y = f(x)$

We start Newton's method with an approximate value for x, say x_1, that is near a root. We can determine the corresponding value of y by the equation $y_1 = f(x_1)$. This will represent a point on the curve that is not, in general, a root. A tangent to $f(x)$ is now constructed at this point on the curve. The tangent is extended until it intersects the x-axis. The

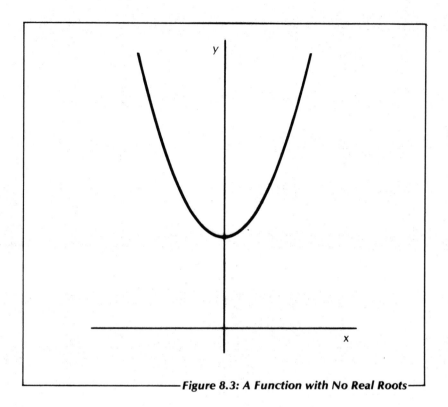

Figure 8.3: A Function with No Real Roots

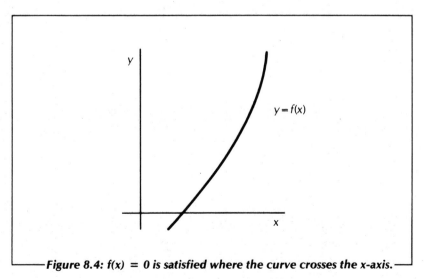

Figure 8.4: f(x) = 0 is satisfied where the curve crosses the x-axis.

next approximation, x_2, is at this intersection on the x-axis, as illustrated in Figure 8.5. Notice that in this example, the second approximation, x_2, is closer to the root than the first approximation, x_1. Thus, we have refined our original approximation.

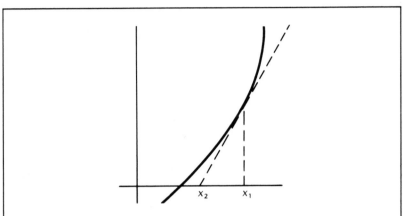

Figure 8.5: The tangent crosses the x-axis closer to the root than the original approximation for x.

The process is now repeated. The function is evaluated at $x = x_2$ to obtain the corresponding value of y, $y_2 = f(x_2)$. The value of y_2 is smaller than the value of y_1, indicating that we are closer to the root. A tangent is again constructed, this time at the point $(x_2, f(x_2))$. The intersection of the new tangent with the x-axis gives the value of x_3, the third approximation of x. We continue in this way, improving the value of x until we are as close to the actual root as we want.

Let us go back and review the first step in more detail. The initial approximation, x_1, gives rise to $y_1 = f(x_1)$. The tangent constructed at y_1 has a slope of:

$$f'(x_1) = \frac{y_1}{x_1 - x_2} \qquad (2)$$

Because $y_1 = f(x_1)$, Equation 2 can be expressed as:

$$x_2 = x_1 - \frac{f(x_1)}{f'(x_1)} \qquad (3)$$

or more generally as:

$$x_{i+1} = x_i - \frac{f(x_i)}{f'(x_i)} \qquad (4)$$

where x_i is the ith approximation. Equation 4 is the usual form of Newton's method. It can be an ideal technique for finding a desired root for a real-life equation.

There are potential problems with the use of Newton's method, which we will consider later in this chapter. But the equations that deal with the behavior of real things typically have only one meaningful root. The other roots of such equations will usually be negative, zero, or complex. Furthermore, the approximate value of the answer may be known. For example, the ideal-gas law can provide a first approximation to a more complicated equation of state.

Now that we have arrived at an equation for the general form of Newton's method, developing our program will be relatively easy.

BASIC PROGRAM: A FIRST ATTEMPT AT NEWTON'S METHOD

We will implement Newton's method for a simple problem, one for which we already know the answer. The equation we will solve is:

$$x^2 = 2$$

or

$$x^2 - 2 = 0 \tag{5}$$

The positive solution to this equation is, of course, the square root of 2. First, we define the function:

$$y = f(x) = x^2 - 2 \tag{6}$$

and its derivative:

$$\frac{dy}{dx} = f'(x) = 2x \tag{7}$$

Our first attempt at a Newton's method program is shown in Figure 8.6. The algorithm itself begins at line 8000. A separate routine, starting at line 8400, is used for the calculation of the function $f(x)$ (Equation 6) and its derivative $f'(x)$ (Equation 7). The first approximation is established at the beginning of the main program. Then the Newton's method subroutine is called to find the solution and to print out the answer.

```
10   REM Newton's method, version 1, Apr 19, 81
11   REM identifiers
13   REM     D6      DX          delta x
15   REM     F       FX          function
16   REM     F1      DFX         derivative of function
```

Figure 8.6: Newton's Method, Version One

```
21   REM     T1      TOL          tolerance
22   REM end of identifiers
23   REM
30   T1 = .000001
40   X = 2
50   GOSUB 8000
60   PRINT
70   PRINT "The  solution  is  "; X
80   GOTO 9999
8000   REM start of Newton's method
8010   REM
8020   REM
8030   X1 = X
8040   GOSUB 8400
8050   REM
8060   REM
8070   REM
8080   REM
8090   D6 = F / F1
8100   X = X1 − D6
8110   PRINT "X = "; X1; ",  fx = "; F; ",  dfx = "; F1
8120   IF (ABS(D6) >= ABS(T1 * X)) THEN 8030
8130   REM
8140   REM
8150   REM
8160   RETURN : REM from Newton's method
8400   F = X * X − 2
8410   F1 = 2 * X
8420   RETURN
9999   END
```

Figure 8.6: Newton's Method, Version One (cont.)

As a matter of good programming practice, subroutines that are used to calculate values should not print intermediate results. But in this particular case, it is instructive to observe the successive values of x, $f(x)$ and $f'(x)$ during the convergence process. Consequently, there is a print statement located at line 8110 for this purpose. You may want to remove this print statement when the program is working properly.

Tolerance

There are a few other matters we should consider at this time. Successive approximations are provided by a loop in the Newton subroutine. This loop continues until two successive values are within the desired tolerance. We are not interested in whether the difference (D6) between two successive approximations has a negative or a positive value. We are only concerned with the magnitude. For this reason, we must be careful to take the absolute value of the comparison.

Furthermore, we are not interested in the actual difference, but only the relative difference. Suppose, for example, that we want our answers to be accurate to one part in a million, that is, one part in 10^6. If a particular solution has a value of unity, then two successive values must be closer than 10^{-6}. If, however, the solution itself has a value of 10^{-6}, then two successive values must differ by no more than 10^{-12}. We therefore choose a relative criterion rather than an absolute criterion for termination of the iteration process.

Generalizing Procedure Calls

Another matter we should consider is the relationship of the Newton's method subroutine to the subroutine it calls for evaluation of the function and its derivative. The Newton's method subroutine gives directions for carrying out the operation described by Equation 4. It is independent of the actual function it is operating on. For this reason, the subroutine that calculates the function and its derivative should be an entirely separate entity.

If two or more different equations are to be solved in the same program, we will have to provide additional, separate copies of the Newton's method subroutine, one for each equation. Another approach would be to use an **ON/GOSUB** construction. This approach would select one equation on the first call, a second equation on the next call, and so on.

Running the Program

Type up the program shown in Figure 8.6 and try it out. The first approximation to the square root of 2 is chosen to be 2. When the program is executed, it should produce the square root of 2 after several iterations. The results will look something like Figure 8.7.

```
X =   2 , fx =   2 , dfx =   4
X =   1.5 , fx =   .25 , dfx =   3
X =   1.41667 , fx =   6.94442E-03 , dfx =   2.83333
X =   1.41422 , fx =   5.96046E-06 , dfx =   2.82843
X =   1.41421 , fx =  -1.19209E-07 , dfx =   2.82843

The solution is  1.41421
```

Figure 8.7: Output: The Positive Root of f(x) = x² − 2

In the following sections we will make several small changes to refine this program. The first change will allow us to input different first-approximation values for the root. This facility is important for studying equations that have more than one root.

Adding User Input for the First Approximation

When the first version of Newton's method is working properly, you can begin to add new features. Make a duplicate copy of the first version. Alter the main program so that it looks like Figure 8.8.

```
10    REM Newton's method, version 2, Apr 21, 81
11    REM identifiers
13    REM        D6      DX          delta x
15    REM        F       FX          function
16    REM        F1      DFX         derivative of function
21    REM        T1      TOL         tolerance
22    REM end of identifiers
23    REM
30    T1 = .000001
40    REM
50    REM
60    REM
70    REM
80    INPUT "First guess "; X
90    IF (X < −19) THEN 9999
100   GOSUB 8000
110   PRINT
```

Figure 8.8: The Main Program for Version Two

```
120    PRINT "The  solution  is  "; X
130    GOTO 80
8000   REM start of Newton's method
       (Continue with lines 8010–8410 of Figure 8.6.)
8420   RETURN
9999   END
```

Figure 8.8: The Main Program for Version Two (cont.)

Run the new version to try it out. For the first version, the value of 2 was used as the first guess to Newton's method. The second version is more sophisticated than the first. The user is asked to input the first guess. The successive approximations, along with the function and its derivative, are displayed as before. At the conclusion of the task, the program begins again and the user is asked to input another first approximation.

Running the Program to Find the Second Root

Start with the value of 2; the results should be the same as for the first version. Then, for the second cycle, try the value of 1. This first approximation is on the other side of the root, but the square root of 2 should again be found in relatively few steps. Try the value of −2 for the third cycle. Notice that the process converges on a different root this time. There are, of course, two solutions to the equation:

$$x^2 - 2 = 0$$

We found the other root this time by giving a negative first approximation.

Investigate what happens when the first guess is near the midpoint of the two roots. Try a first guess of 0.0001. In this case, the process takes quite a few steps to produce the answer. Finally, try a first guess of zero. The curve

$$y = f(x)$$

has zero slope at this point. Consequently, one of two things can happen. Either a floating-point divide error can occur or the program can loop indefinitely. This problem will be corrected in the next version.

A Test for Zero Slope

When the derivative, or slope, of our function is zero, the final term in Equation 4 becomes infinite. We are looking for the point where the slope crosses the x-axis. But it can be seen from Figure 8.9 that the tangent to the curve at $x = 0$ is parallel to the x-axis. Consequently, the two lines do not intersect.

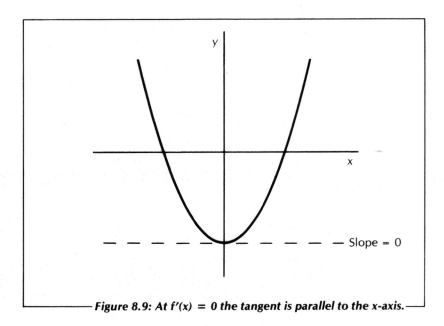

Figure 8.9: At f'(x) = 0 the tangent is parallel to the x-axis.

We will now add some instructions for testing the slope to our Newton's method program. We will define a small number such as:

$$H2 = 1E-15;$$

Then, the slope (F1) can be tested with the statement:

IF (ABS(F1) > H2) **THEN** . . .

One problem with this approach, however, is that the value assigned to H2 must be consistent with the particular version of BASIC being used. That is, the value of H2 may have to be chosen carefully.

We will also define two variables, F0% (false) and T0% (true), as logical indicators. Then we can set an error flag, E1%, to a value of TRUE if there is a problem. The main program can test this flag after each return from the Newton's method procedure. If the flag is not set, then there is no error and the solution is printed. Alter the Newton's method procedure so that it looks like Figure 8.10. Then try out the new version.

```
10      REM Newton's method, version 3, Apr 21, 81
11      REM identifiers
13      REM       D6        DX          delta x
14      REM       E1%       ERMES%      error flag
15      REM       F         FX          function
16      REM       F1        DFX         derivative of function
17      REM       F0%       FALSE%      zero
18      REM       H2        SMALL       small number
20      REM       T0%       TRUE%       not false
21      REM       T1        TOL         tolerance
22      REM end of identifiers
23      REM
30      T1 = .000001
40      H2 = 1E−15
50      REM
60      F0% = 0
70      T0% = NOT F0%
80      INPUT "First  guess "; X
90      IF (X < −19) THEN 9999
100     GOSUB 8000
110     PRINT
120     IF (E1% = F0%) THEN PRINT "The  solution  is "; X
130     GOTO 80
8000    REM start of Newton's method
8010    E1% = F0%
8020    REM
8030    X1 = X
8040    GOSUB 8400
8050    IF (ABS(F1) > H2) THEN 8090
8060    PRINT "ERROR—slope is zero"
8070    E1% = T0%
8080    GOTO 8160
8090    D6 = F / F1
8100    X = X1 − D6
8110    PRINT "X = "; X1;", fx = ";F;", dfx = ";F1
```

Figure 8.10: Newton's Method with a Test for Zero Slope

```
8120    IF (ABS(D6) >= ABS(T1 * X)) THEN 8030
8130    REM
8140    REM
8150    REM
8160    RETURN : REM from Newton's method
8400    F = X * X - 2
8410    F1 = 2 * X
8420    RETURN
9999    END
```

Figure 8.10: Newton's Method with a Test for Zero Slope (cont.)

Running the Program with the Slope Test

Give initial values 2, 1, and -1 as before to see that the program behaves properly. Then, give an initial value of zero. This input caused a floating-point divide check in the previous version; now, your BASIC program should handle this input with no difficulty. The program should print the appropriate error message, then request another first approximation. The program can be aborted by entering a value that is less than -19.

Finally, our next task is to make sure the program will print an error message and terminate after an appropriate number of iterations if approximations do not converge on a root.

Failure to Converge

Sometimes, Newton's method will not converge on a root after a reasonable number of iterations. One possibility is that successive approximations are oscillating around a complex root as illustrated in Figure 8.11.

The first approximation, x_1, gives rise to the second value, x_2. But x_2 then produces the original value of x_1 for the third approximation. In this case, the process will never terminate.

Another possibility is that an approximation is very far from a root. This can occur even though the first guess appears to be close to a root. We can observe this behavior with our previous version of Newton's method if we run the program again and give a first approximation of 0.00001. This will produce a second value of 20,000, which is quite an overshoot. Each successive approximation will now be about one-half of the previous value. A solution will eventually be obtained in this case. But it will take more than 20 iterations before the values converge.

We can guard against lack of convergence by adding a loop counter.

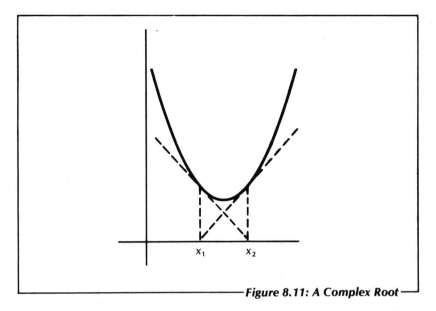

Figure 8.11: A Complex Root

We can then abort the process if convergence does not occur after, say, 20 iterations. This will take care of the case of oscillation, as well as the case of an approximation that is very far from a root.

Alter the Newton's method procedure so that it looks like Figure 8.12. Run the new version to try it out. Give an initial approximation of 500000. This first guess is so far from the root that it will take many cycles to converge. The iteration process should stop after 20 loops and the appropriate error message will be displayed. Next, try a first guess of 2 to make sure that everything is still all right.

```
10   REM Newton's method, version 4, Apr 21, 81
11   REM identifiers
13   REM    D6       DX         delta x
14   REM    E1%      ERMES%     error flag
15   REM    F        FX         function
16   REM    F1       DFX        derivative of function
17   REM    F0%      FALSE%     zero
18   REM    H2       SMALL      small number
19   REM    M5%      MAXL%      maximum loops
20   REM    T0%      TRUE%      not false
```

Figure 8.12: Newton's Method with a Loop Counter

```
  21    REM      T1        TOL           tolerance
  22    REM end of identifiers
  23    REM
  30    T1 = .000001
  40    H2 = 1E−15
  50    M5% = 20
  60    F0% = 0
  70    T0% = NOT F0%
  80    INPUT "First guess "; X
  90    IF (X < −19) THEN 9999
 100    GOSUB 8000
 110    PRINT
 120    IF (E1% = F0%) THEN PRINT "The  solution  is "; X
 130    GOTO 80
8000    REM start of Newton's method
8010    E1% = F0%
8020    FOR I% = 1 TO M5%
8030       X1 = X
8040       GOSUB 8400
8050       IF (ABS(F1) > H2) THEN 8090
8060       PRINT "ERROR—slope  is  zero"
8070       E1% = T0%
8080       GOTO 8160
8090       D6 = F / F1
8100       X = X1 − D6
8110       PRINT "X  =  "; X1; ",  fx  =  "; F; ",  dfx  =  "; F1
8120       IF (ABS(D6) <= ABS(T1 * X)) THEN 8160
8130       NEXT I%
8140       PRINT "ERROR—no  convergence  in "; M5%; "  tries"
8150    E1% = T0%
8160    RETURN : REM from Newton's method
8400    F = X * X − 2
8410    F1 = 2 * X
8420    RETURN
9999    END
```

Figure 8.12: Newton's Method With a Loop Counter (cont.)

We have developed and refined our program using a simple, predictable function. Now we are ready to use our program to find the roots of some more difficult functions.

BASIC PROGRAM: SOLVING OTHER EQUATIONS

Consider the nonlinear equation:

$$e^x = 4x$$

This equation, unlike our previous one, cannot be solved explicitly for x. Consequently, it is a suitable candidate for our Newton's method program. The corresponding function and its derivative are:

$$f(x) = e^x - 4x$$

and

$$f'(x) = e^x - 4$$

Replace lines 8400-8420 with the lines shown in Figure 8.13. Notice that the variable E is introduced so that the exponent of X will not have to be calculated twice. Recall that

$$\frac{de^x}{dx} = e^x$$

```
8400   E = EXP(X)
8410   F = E − 4 * X
8420   F1 = E − 4
8430   RETURN
```

Figure 8.13: Subroutine for Evaluating $e^x = 4x$

Run the new version and input a first guess of 4. The program should converge on a value of 2.153. This is one of the two roots. The other root can be found by giving a first guess of 0.1. The result, in this case, will be a value of 0.357.

Because our next equation involves the SIN function, its implementation will require one of two versions, depending on the BASIC in use.

A Function with Many Roots

Let us explore the several roots of the equation:

$$\sin(x) = \frac{x}{10}$$

Change the program starting with line 8400 so that it looks like Figure 8.14.

```
8400   REM
8410   F = SIN(X) — 0.1 * X
8420   F1 = COS(X) — 0.1
8430   RETURN
9999   END
```

Figure 8.14: The Solution of sin(x) = x / 10

Zero is one of the roots of the new equation. As we approach this root, we will need to take the SIN of numbers that are closer and closer to zero. Unfortunately, as we discovered in Chapter 1, several BASICs contain an error that will cause trouble in this case. The problem is that an incorrect value may be returned for the SIN.

If you have not tested your SIN function with the program given in Chapter 1, you should do so at this time. If the results of the test indicate a problem with your SIN, then you should use the subroutine given in Figure 8.15 rather than the one in Figure 8.14. The alternate approach is to inspect the argument of the SIN. If the number is closer to zero than a value such as 0.000001, then the value of the argument is returned. Otherwise, the built-in SIN function is called.

```
8400   F = X — 0.1 * X
8410   IF (ABS(X) > 0.000001) THEN F = SIN(X) — 0.1 * X
8420   F1 = COS(X) — 0.1
8430   RETURN
9999   END
```

Figure 8.15: Alternate Implementation for sin(x) = x / 10

Run this latest version to try it out. Give a first estimate of 1.0. This should converge on the root at zero. Then, try a first guess of −1.0. This too should converge on the root at zero. There are several other roots to this equation. The table in Figure 8.16 gives several first approximations and the corresponding roots. Try out each one to verify that your Newton's method is working properly.

1st approximation	root
− 1	0
1	0
4	2.852..
4.3	7.068..
4.5	0
4.7	− 8.423..
5	− 2.852..
6	7.068..
9	8.423..

Figure 8.16: The Roots of sin(x) − x / 10

BASIC PROGRAM: THE VAPOR PRESSURE EQUATION

We are now ready to solve the vapor pressure equation that was introduced at the beginning of this chapter. We will write our function and its derivative as:

$$f(T) = A + \frac{B}{T} + C \ln T - \ln P$$

$$df(T) = \frac{-B}{T^2} + \frac{C}{T}$$

Remember that A, B, C, and P are constants. Make a copy of the Newton program and add the lines shown in Figure 8.17. With this function we will find the temperature that corresponds to a vapor pressure of 0.01 atmospheres.

Run the new version to try it out. Give a first guess of 500 degrees and the program will converge to a temperature of 1416 degrees in about seven steps.

```
  32   A = 18.19
  34   B = −23180
  36   C = −.8858
  38   P = LOG(0.01)
 . . .
8400   F = A + B / X + C * LOG(X) − P
8410   F1 = −B / (X * X) + C / X
8420   RETURN
```

Figure 8.17: Solution of the Vapor Pressure Equation

SUMMARY

Using a familiar function at first, we saw how easy it was to write, and then improve upon, a BASIC program implementing Newton's method for finding the roots of an equation. We then used our program to solve some functions that are more complex. These included exponential and trigonometric functions. Finally, we solved the equation of an actual scientific application.

EXERCISES

8-1: *The degree of dissociation, x, of hydrogen sulfide gas can be described by the equation:*

$$(1 - PK^2)\, x^3 - 3\,x + 2 = 0$$

where K is the equilibrium constant, P is the total pressure in atmospheres. For a temperature of 2000° Kelvin, K = 0.608. Use Newton's method to find the degree of dissociation when the pressure is one atmosphere. Since x only has meaning over the range 0-1, begin with an approximation of 0.5. Explore the consequences of initial guesses of zero, of unity, and of 1.2.

Answer: x = 0.758

8-2: *Van der Waals' equation of state is:*

$$(P + A/V^2)(V - B) = RT$$

where P is the pressure, V is the molar volume, T is the temperature and R is the gas constant. Coefficient A is a measure of the bonding force and coefficient B is a volume correction. When the pressure is stated in atmospheres, the volume in liters and the temperature in degrees Kelvin, R has a value of 0.082 liter-atm/deg mole. Coefficient A is in units of pressure times molar volume squared and B has units of molar volume. The coefficients A and B have been extensively tabulated for common gases. For example, the values for toluene are:

$$A = 24.06 \text{ liter}^2 \text{ atm/mole}^2$$

and

$$B = 0.1463 \text{ liter/mole}$$

Use Newton's method to solve the Van der Waals equation. Find the molar volume of toluene at the boiling point of 110° C (383° K) and a pressure of 1 atm. The variables A, B, V, P, R, and T can all be used in the BASIC program. Convert X to V at line 8400. Then define the function f(x) and its derivative:

```
8400 V = X
8410 F = V − R*T/P − B + A/(P*V) − A*B/(V*V)
8420 F1 = 1.0 − A/(P*V*V) + A*B/(V*V*V)
```

The values of A, B, T, P and R are defined near the beginning of the program. Change line 80 so that the program calculates the first approximation from the ideal gas law: V = RT/P. Change line 130 to STOP. Run the program to find the Van der Waals volume.

Answer: 30.8 liters

8-3: *Use Newton's method to solve the Van der Waals equation as in the previous problem, but input the temperature and pressure from the keyboard. Print both the ideal gas volume and the Van der Waals volume.*

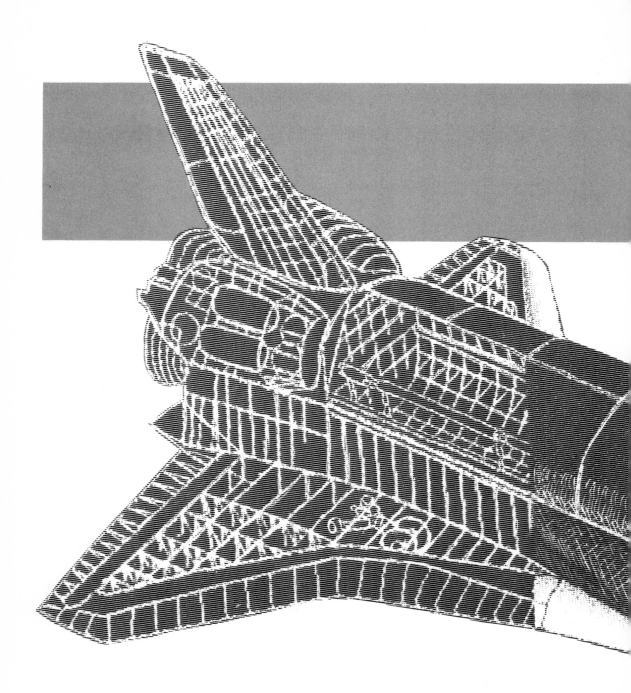

CHAPTER **9**

Numerical Integration

INTRODUCTION

In this chapter, we will develop three different methods for carrying out numerical integration; that is, for determining the area beneath a curve between two given values of the independent variable. After writing BASIC programs for each of the methods, we will judge how efficiently each method computes the area. Each of the methods uses progressively smaller ''panels'' of measurable area to divide up the area beneath the curve. The summation of the areas of these panels then provides an approximation of the total area.

In the trapezoidal rule method, these panels are topped by straight-line secants to the curve, whereas Simpson's method tops each panel with a parabolic curve of its own. Both of these methods can be improved with an abbreviated Taylor series expansion to compute the error—a method referred to as *end correction*. The third integration method we will study, the Romberg method, is the most complex; it constructs a matrix of interpolations to arrive at the total area. One of the functions we will try on both the Simpson and Romberg methods is related to the normal distribution function, which we will be returning to in Chapter 11. Finally, we will explore a method for integrating a function that approaches infinity at one of its limits.

Let us begin with a brief description of the definite integral and the methods of evaluating it.

THE DEFINITE INTEGRAL

The evaluation of the *definite integral*:

$$\int_a^b f(x)dx = F(b) - F(a)$$

where $F'(x) = dF(x)/dx = f(x)$, can be interpreted as the area under the curve of the function $f(x)$ from the limit a to the limit b, as illustrated in Figure 9.1.

The actual integration can be straightforward for certain functions, but very difficult for others. For example, the power series:

$$1 + 2x + 3x^2$$

can be integrated term by term and then evaluated between the limits of a and b to give:

$$b + b^2 + b^3 - a - a^2 - a^3$$

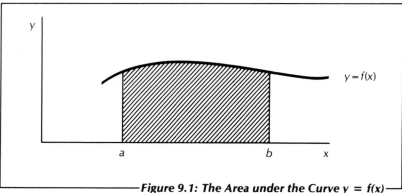

$$y = f(x)$$

Figure 9.1: The Area under the Curve $y = f(x)$

More complicated functions can sometimes be integrated by reference to integration tables found in math handbooks. Other techniques, such as *integration by parts*, can sometimes be used. Another possibility is to replace the original function with an infinite series or an *asymptotic expansion*. The integration is then carried out on the replacement function.

Sometimes the integrand is not a proper function at all, but simply a collection of experimental data. In such a case, it may be possible to fit the experimental data to a function that can be integrated. Nevertheless, there will be times when it will be impossible or very difficult to evaluate the integrand analytically. But if the limits are known, then an approximation method may provide an acceptable solution. This approach is known as *numerical integration*.

Several different methods are commonly used for numerical integration. These typically involve the substitution of an easily integrated function for the original function. The new function may be a polynomial such as a straight line or parabola, or it may consist of transcendental functions such as sines and cosines. The accuracy of the resulting calculation depends on how well the substituted function approximates the original function.

In the next section we will describe both the general method of function substitution, and a specific, simple implementation of the method— the use of straight-line functions.

THE TRAPEZOIDAL RULE

One of the simplest methods of numerical integration is known as the *trapezoidal rule*. In this method, the original function is approximated by a set of straight lines. The region to be integrated is divided into uniformly spaced sections or panels. The panel width, Δx, is:

$$\Delta x = \frac{b - a}{n}$$

where n is the number of panels and a and b are the integration limits. If the entire integral is fitted with a single straight line, that is, if there is only one panel (as illustrated in Figure 9.2) then the calculated area is:

$$\frac{(b - a)[f(a) + f(b)]}{2}$$

In this formula, $f(a)$ is the value of the function at the left limit and $f(b)$ is the value at the right limit.

For the more general case, the area is divided into n panels. An interior panel is bounded on the left by a vertical line at x_i, and on the right by a vertical line at $x_i + \Delta x$. The lower edge of the panel is marked by the x axis. The original curve at the top of the panel is replaced by a straight line that in general will not have a zero slope (that is, it will not be parallel to the x-axis). The resulting panel thus has the shape of a trapezoid, giving rise to the name of the method.

The panels can be numbered from 1 to n, but actually we are interested in evaluating the function at the left and right edges of each panel. There are $n - 1$ interior edges for the n panels, in addition to the right and left boundaries of the integral. Consequently, there will be $n + 1$ edges, which can be numbered from 0 through n.

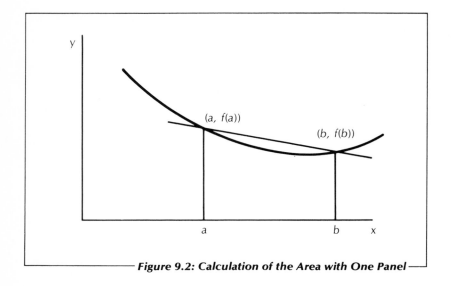

Figure 9.2: Calculation of the Area with One Panel

The area of the first panel is:

$$\frac{[f(0) + f(1)]\Delta x}{2}$$

or

$$\frac{[f(a) + f(1)]\Delta x}{2}$$

and the area of the last panel is:

$$\frac{[f(n-1) + f(n)]\Delta x}{2}$$

or

$$\frac{[f(n-1) + f(b)]\Delta x}{2}$$

The area of the ith panel is:

$$\frac{[f(i-1) + f(i)]\Delta x}{2}$$

where $f(i-1)$ is the value of the function on the left side of the ith panel and $f(i)$ is the value of the function on the right side of the panel. The desired integral is the sum of the areas of all of the panels. Thus the total area can be calculated by summing the areas of all of the panels according to the expression:

$$\{[f(a) + f(1)] + [f(1) + f(2)] + [f(2) + f(3)] + \ldots$$
$$+ [f(n-2) + f(n-1)] + [f(n-1) + f(b)]\} \frac{\Delta x}{2}$$

The right edge of the first panel is also the left edge of the second panel, and the left edge of the last panel is also the right edge of the next to last panel. The edges of all of the other panels are also common to two panels. The formula for integration by the trapezoidal rule can therefore be simplified as:

$$[f(a) + 2f(1) + 2f(2) + \ldots$$
$$+ 2f(n-2) + 2f(n-1) + f(b)] \frac{\Delta x}{2}$$

Our first computer program for this method will allow us to experiment with the number of panels we use to divide up the total area.

BASIC PROGRAM: THE TRAPEZOIDAL RULE WITH USER INPUT FOR THE NUMBER OF PANELS

The area calculated from the trapezoidal method more closely approaches the actual value as the number of panels becomes larger and larger. This can be demonstrated with the program given in Figure 9.3. Type up the program and run it. The main program will ask for the number of sections into which the area is to be divided. The resulting calculated value will then be based on the given number of panels.

```
10   REM Integration by trapezoidal rule, Jul 17, 81
11   REM
12   REM identifiers
14   REM      D7       DELTA       panel width
15   REM      E6       ESUM        end sum
16   REM      F2       FX          function to integrate
17   REM      L7       LOWER       lower limit
18   REM      P5%      PIECES%     number of panels
19   REM      P6       PSUM        panel sum
20   REM      S3       SUM
21   REM      U2       UPPER       upper limit
```

Figure 9.3: The Trapezoidal Rule

```
 22    REM end of identifiers
 23    REM
 30    DEF FNF2(X) = 1 / X
 40    INPUT "How many sections"; P5%
 50    IF (P5% < 0) THEN 9999
 60    L7 = 1
 70    U2 = 9
 80    GOSUB 2000
 90    PRINT " area = "; S3
100    GOTO 40
2000   REM integration by trapezoidal rule
2010   D7 = (U2 − L7) / P5%
2020   E6 = FNF2(L7) + FNF2(U2)
2030   P6 = 0
2040   FOR I% = 1 TO P5% − 1
2050     X = L7 + I% * D7
2060     P6 = P6 + FNF2(X)
2070   NEXT I%
2080   S3 = (E6 + 2 * P6) * D7 / 2
2090   RETURN
9999   END
```

Figure 9.3: The Trapezoidal Rule (cont.)

The algorithm for integration by the trapezoidal rule begins at line 2000. The function to be integrated:

$$\int_1^9 \frac{dx}{x}$$

is defined as function F2 at line 30. Since this function can be readily integrated, we can compare the exact value of the integral to the value calculated by the trapezoidal method. The value of the integral in this case is the natural logarithm of 9, or 2.197225.

We will now look at a more sophisticated version of this program.

BASIC PROGRAM: AN IMPROVED TRAPEZOIDAL RULE

We can improve our trapezoidal program in two ways. First, we can change the subroutine starting at line 2000 so that it will automatically divide the original area into more and more pieces. Second, we can avoid much calculation at each step by using the results from the previous step. Alter the first trapezoidal program so that it looks like the one given in Figure 9.4 and then run it.

```
10    REM Integration by trapezoidal rule, Apr 19, 81
11    REM
12    REM      identifiers
14    REM      D7     DELTA      panel width
15    REM      E6     ESUM       end sum
16    REM      F2     FX         function to integrate
17    REM      H3     SUM1
18    REM      L7     LOWER      lower limit
19    REM      P5%    PIECES%    number of panels
20    REM      P6     PSUM       panel sum
21    REM      S3     SUM
22    REM      T1     TOL        tolerance
23    REM      U2     UPPER      upper limit
24    REM end of identifiers
25    REM
26    REM
30    DEF FNF2(X) = 1 / X
40    REM
50    T1 = .00001
60    L7 = 1
70    U2 = 9
80    GOSUB 2000
90    PRINT
100   PRINT " area  = "; S3
110   GOTO 9999
2000  REM integration by trapezoidal rule
2010  P5% = 1
2020  D7 = (U2 − L7) / P5%
2030  E6 = FNF2(L7) + FNF2(U2)
2040  REM
```

Figure 9.4: An Improved Trapezoidal Method

```
2050    S3 = E6 * D7 / 2
2060    PRINT "       1 "; S3
2070    P6 = 0
2080    P5% = P5% * 2
2090    H3 = S3
2100    D7 = (U2 − L7) / P5%
2110    FOR I% = 1 TO P5% / 2
2120       X = L7 + D7 * (2 * I% − 1)
2130       P6 = P6 + FNF2(X)
2140    NEXT I%
2150    REM
2160    S3 = (E6 + 2 * P6) * D7 / 2
2170    PRINT USING " #####   #.#####"; P5%, S3
2180    IF (ABS(S3 − H3) > ABS(T1 * S3)) THEN 2080
2190    RETURN
9999    END
```

Figure 9.4: An Improved Trapezoidal Method (cont.)

Running the Program

For the first calculation, the entire area is taken as a single panel. The number of panels is then doubled to two and the new area is calculated. The process is continued with a doubling of the number of panels at each step. As the number of panels increases, so does the accuracy of the result and, of course, the length of the computation time. The number of panels and the corresponding calculated areas are displayed at each step. The output should look like Figure 9.5.

```
        1  4.44444
        2  3.02222
        4  2.46349
        8  2.27341
       16  2.21733
       32  2.20234
       64  2.19851
      128  2.19755
      256  2.19731
      512  2.19725
     1024  2.19723

     area =   2.19723
```

Figure 9.5: Integration by the Trapezoidal Method (The number of panels and the corresponding areas are given.)

It can be seen from Figure 9.5 that the calculated value more closely approaches the correct value as the number of panels is increased. The process terminates when two successive values are within the desired tolerance. The tolerance (T1) is set to a value of 10^{-5} at line 50. You may want to change this value to correspond to the precision of your BASIC. You may also want to remove the statement at line 2170 that prints the successive values during the iteration process.

With this version, the number of panels at each step is twice the number used for the previous step. Notice that it is not necessary to recalculate all of the panel heights at each step. Suppose, for example, that four panels are used for a particular step. The next step will then use eight panels. Since all of the panel heights from the previous step are common to the present step, it is faster to save the sum from each step and use it for the next step, rather than recalculate all the heights. Only the heights midway between the previous points need to be computed. These are then added to the previous sum of the interior values. The new sum is multiplied by 2 and added to the values at each end of the function. This result is then multiplied by one-half the panel width to obtain the area.

Although this improved version runs faster than the first version, it suffers from the same errors. We are trying to fit a general curve with a set of straight lines. If the original curve is not a straight line, many steps may be needed for convergence.

In our final version of the trapezoidal rule method we will include a calculation of end correction in our program.

End Correction

We can further improve the trapezoidal method by including correction factors obtained from a Taylor series expansion. The error terms contain second derivatives of the function. Fortunately, however, simplified correction terms can be used instead. The terms for the interior points vanish, leaving only the terms at the upper and lower limits of the function. Thus the major error-correction term requires a derivative of the function at the two limits, x_0 and x_n. The additional quantity to be included in the sum is:

$$\frac{[f'(b) - f'(a)](\Delta x)^2}{12}$$

where $f'(b)$ is the value of the slope at b and $f'(a)$ is the value of the slope at a. The resulting value is subtracted from the regular trapezoidal sum. The method is referred to as integration by the trapezoidal rule with end correction.

BASIC PROGRAM: TRAPEZOIDAL RULE WITH END CORRECTION

Make a copy of the previous program. Add the derivative function at line 40. Change the subroutine at line 2000 according to Figure 9.6.

```
   10   REM Integration by trapezoidal rule, Apr 19, 81
   11   REM
   12   REM identifiers
   14   REM     D7      DELTA       panel width
   15   REM     D8      DFX         derivative
   16   REM     E6      ESUM        end sum
   17   REM     E7      ECOR        end correction
   18   REM     F2      FX          function to integrate
   19   REM     H3      SUM1
   20   REM     L7      LOWER       lower limit
   21   REM     P5%     PIECES%     number of panels
   22   REM     P6      PSUM        panel sum
   23   REM     S3      SUM
   24   REM     T1      TOL         tolerance
   25   REM     U2      UPPER       upper limit
   26   REM end of identifiers
   27   REM
   30   DEF FNF2(X) = 1 / X
   40   DEF FND8(X) = −1 / (X * X)
   50   T1 = .00001
   60   L7 = 1
   70   U2 = 9
   80   GOSUB 2000
   90   PRINT
  100   PRINT " area = "; S3
  110   GOTO 9999
 2000   REM integration by trapezoidal rule
 2010   P5% = 1
 2020   D7 = (U2 − L7) / P5%
 2030   E6 = FNF2(L7) + FNF2(U2)
 2040   E7 = (FND8(U2) − FND8(L7)) / 12
```

Figure 9.6: Trapezoidal Rule with End Correction

```
2050    S3 = E6 * D7 / 2
2060    PRINT "      1 "; S3
2070    P6 = 0
2080    P5% = P5% * 2
2090    H3 = S3
2100    D7 = (U2 − L7) / P5%
2110    FOR I% = 1 TO P5% / 2
2120       X = L7 + D7 * (2 * I% − 1)
2130       P6 = P6 + FNF2(X)
2140    NEXT I%
2150    S3 = (E6 + 2 * P6) * D7 / 2
2160    S3 = S3 − D7 * D7 * E7
2170    PRINT USING " #####   #.#####"; P5%, S3
2180    IF (ABS(S3 − H3) > ABS(T1 * S3)) THEN 2080
2190    RETURN
9999    END
```

Figure 9.6: Trapezoidal Rule with End Correction (cont.)

Running the Program

Run the program and compare the results to Figure 9.7. Notice how much faster the process converges in this case. Also notice that during the iteration process, the sum passed the correct value, then reversed its direction. It should be realized, however, that it will not always be possible to use the end correction technique with the trapezoidal method. This will be the case if the slope at one limit or the other is infinite or becomes very large. (We will discuss this case at the end of this chapter.) Also, if the function has a zero slope at the limits, the end-correction term will be zero.

```
        1  4.44444
        2  1.70535
        4  2.13427
        8  2.19111
       16  2.19675
       32  2.19719
       64  2.19722
      128  2.19723

     area =   2.19723
```

Figure 9.7: Trapezoidal Integration with End Correction

The trapezoidal rule method replaces the function with a first-order polynomial (a straight line). In many cases we can expect a second-order polynomial (a parabola) to be a better substitute; this is the idea of the method we will implement in our next program. We will examine three different versions of this program—first with the same function we ran on the trapezoidal rule program, then with an exponential and a sine function.

BASIC PROGRAM: SIMPSON'S INTEGRATION METHOD

Simpson's method of numerical integration is similar to the trapezoidal method except that the original curve is replaced by a set of parabolas rather than straight lines. Since a parabola can be defined by a minimum of three points, we must divide the original area into an even number of panels. Each parabola is fitted to the tops of two adjacent panels. In general, we expect parabolas to produce a better fit than straight lines. That is, we should need fewer panels to obtain satisfactory convergence. The formula is:

$$(f_0 + f_n + 4 \sum_{\substack{j=1 \\ j \text{ odd}}}^{n-1} f_j + 2 \sum_{\substack{j=2 \\ j \text{ even}}}^{n-2} f_j) \frac{\Delta x}{3}$$

First Run of the Simpson's Rule Program

Type up the program shown in Figure 9.8 and run it. The results should look like Figure 9.9.

10	**REM** Integration by Simpson's rule, Apr 19, 81		
11	**REM**		
12	**REM** identifiers		
14	**REM**	D7	DELTA	panel width
15	**REM**	E6	ESUM	end sum
16	**REM**	E8	EVSUM	even sum
17	**REM**	F2	FX	function to integrate
18	**REM**	H3	SUM1	
19	**REM**	L7	LOWER	lower limit
20	**REM**	O2	ODSUM	odd sum
21	**REM**	P5%	PIECES%	number of panels
22	**REM**	S3	SUM	
23	**REM**	T1	TOL	tolerance
24	**REM**	U2	UPPER	upper limit

Figure 9.8: Simpson's Rule

```
  25    REM end of identifiers
  26    REM
  30    DEF FNF2(X) = 1 / X
  40    REM
  50    T1 = .00001
  60    L7 = 1
  70    U2 = 9
  80    GOSUB 2000
  90    PRINT
 100    PRINT " area = "; S3
 110    GOTO 9999
2000    REM integration by Simpson's rule
2010    P5% = 2
2020    D7 = (U2 − L7) / P5%
2030    O2 = FNF2(L7 + D7)
2040    E8 = 0
2050    E6 = FNF2(L7) + FNF2(U2)
2060    REM
2070    S3 = (E6 + 4 * O2) * D7 / 3
2080    PRINT USING " #####   #.#####"; P5%, S3
2090    P5% = P5% * 2
2100    H3 = S3
2110    D7 = (U2 − L7) / P5%
2120    E8 = E8 + O2
2130    O2 = 0
2140    FOR I% = 1 TO P5% / 2
2150       X = L7 + D7 * (2 * I% − 1)
2160       O2 = O2 + FNF2(X)
2170    NEXT I%
2180    S3 = (E6 + 4 * O2 + 2 * E8) * D7 / 3
2190    REM
2200    PRINT USING " #####   #.#####"; P5%, S3
2210    IF (ABS(S3 − H3) > ABS(T1 * S3)) THEN 2090
2220    RETURN
9999    END
```

Figure 9.8: Simpson's Rule (cont.)

```
              2  2.54815
              4  2.27725
              8  2.21005
             16  2.19864
             32  2.19734
             64  2.19723
            128  2.19723

          area =   2.19723
```

Figure 9.9: Integration by Simpson's rule

Compare Figure 9.9 with Figures 9.5 and 9.7. At each point, the value of the integral obtained by Simpson's rule is closer to the correct answer than is the value obtained from the straight trapezoidal approach for the same number of panels. As a result, fewer operations are needed and so convergence occurs more quickly. On the other hand, the convergence is about the same as the trapezoidal method with end correction.

At each step of Simpson's method, the function is evaluated only at the odd positions. The sum for the even positions is obtained from the sum of all of the previous interior positions, both even and odd.

Second Run of the Simpson Program—An Exponential Function

Let us consider a second example of integration by the Simpson method. The following function is related to the normal distribution described in Chapter 2. It is also related to the error function described in Chapter 11. At this time, we will integrate the curve from zero to a value of unity.

Make a copy of the previous program and alter it so that it will find the value of the integral:

$$\int_0^1 e^{-x^2} dx$$

Change function F2 at line 30 so that it looks like:

30 **DEF** FNF2(X) = EXP(−X ∗ X)

Then change the integration limits at lines 60 and 70 so that they read:

60 L7 = 0
70 U2 = 1

Run the new version and compare the output to Figure 9.10. Notice how rapidly the process converges for this function.

```
              2      0.747180
              4      0.746855
              8      0.746826
             16      0.746824

           area  =   0.746824
```

Figure 9.10: Integration of e^{-x^2} by Simpson's Rule

Third Run of the Simpson Program—A Periodic Function

All the functions we have integrated so far exhibit curvature in the same direction throughout the interval. Let us now consider the periodic function:

$$\sin^2 x$$

over the interval of 0 to 4π. This function is always positive; consequently, the integral over any region should produce a positive number. But this function has a zero value at both limits, as well as at the midpoint of the interval and at the quarter points. Consequently, the first area calculated by either the trapezoidal method or the Simpson method will give a zero result. The next approximations, however, will give positive results with Simpson's method.

Make a copy of the previous program to solve this function over the limit of 0 to 4π. Change line 40 to:

    40 P7 = 3.14159

Alter function F2 to look like:

    30 **DEF** FNF2(X) = SIN(X)^2

Set the integration limits to:

    60 L7 = 0
    70 U2 = 4 * P7

There is another factor that must be considered in this case. Since the function has a value of zero at several strategic locations, the calculated area for two panels and for four panels gives a result of zero. Consequently, we must begin the process with four panels. Therefore, change line 2010 to read:

    2010 P5% = 4

Run the program and compare the output to Figure 9.11.

```
                    4  0
                    8  8.37758
                   16  6.28319
                   32  6.28319

                   area =   6.28319
```

Figure 9.11: The Integral of sin²x over 0 to 4π

We will now derive an end-correction formula for Simpson's method.

BASIC PROGRAM:
THE SIMPSON METHOD WITH END CORRECTION

Error correction can be applied to the Simpson method, as it was with the trapezoidal rule. The principal term contains the fourth derivative of the function. In this case, however, it is better to use an approximation requiring only the first derivative. As with the trapezoidal correction, the derivative is only needed at the end points. The Simpson's rule formula with end correction now looks like:

$$\{7(f_0 + f_n) + 14 \sum_{\substack{j=2 \\ j \text{ even}}}^{n-2} f_j + 16 \sum_{\substack{j=1 \\ j \text{ odd}}}^{n-1} f_j + \Delta x[f'(a) - f'(b)]\} \frac{\Delta x}{15}$$

Running the Program with End Correction

Make a copy of the Simpson integration program shown in Figure 9.8. Add the derivative function at line 40 that was used in the trapezoidal method with end correction. Alter the subroutine at 2000 so that it looks like Figure 9.12. Run the program and compare the results to Figure 9.13.

```
10   REM Integration by Simpson's rule, Apr 19, 81
11   REM
12   REM identifiers
14   REM     D7     DELTA     panel width
15   REM     D8     DFX       derivative
16   REM     E6     ESUM      end sum
17   REM     E7     ECOR      end correction
18   REM     E8     EVSUM     even sum
```

Figure 9.12: Simpson's Rule with End Correction

```
19    REM    F2      FX              function to integrate
20    REM    H3      SUM1
21    REM    L7      LOWER           lower limit
22    REM    O2      ODSUM           odd sum
23    REM    P5%     PIECES%         number of panels
24    REM    S3      SUM
25    REM    T1      TOL             tolerance
26    REM    U2      UPPER           upper limit
27    REM end of identifiers
28    REM
30    DEF FNF2(X) = 1 / X
40    DEF FND8(X) = −1 / (X * X)
50    T1 = .00001
60    L7 = 1
70    U2 = 9
80    GOSUB 2000
90    PRINT
100   PRINT " area = "; S3
110   GOTO 9999
2000  REM integration by Simpson's rule
2010  P5% = 2
2020  D7 = (U2 − L7) / P5%
2030  O2 = FNF2(L7 + D7)
2040  E8 = 0
2050  E6 = FNF2(L7) + FNF2(U2)
2060  E7 = FND8(L7) − FND8(U2)
2070  S3 = (E6 + 4 * O2) * D7 / 3
2080  PRINT USING "    ### ##.#####"; P5%, S3
2090  P5% = P5% * 2
2100  H3 = S3
2110  D7 = (U2 − L7) / P5%
2120  E8 = E8 + O2
2130  O2 = 0
```

Figure 9.12: Simpson's Rule with End Correction (cont.)

```
2140    FOR I% = 1 TO P5% / 2
2150        X = L7 + D7 * (2 * I% − 1)
2160        O2 = O2 + FNF2(X)
2170    NEXT I%
2180    S3 = 7 * E6 + 14 * E8 + 16 * O2 + E7 * D7
2190    S3 = (S3 * D7) / 15
2200    PRINT USING " #####   #.#####"; P5%, S3
2210    IF (ABS(S3 − H3) > ABS(T1 * S3)) THEN 2090
2220    RETURN
9999    END
```

Figure 9.12: Simpson's Rule with End Correction (cont.)

```
        2    2.54815
        4    2.16287
        8    2.19490
       16    2.19713
       32    2.19722
       64    2.19722

  area  =    2.19722
```

Figure 9.13. Integration by Simpson's Rule with End Correction

Convergence in this case is the fastest of the four techniques we have used. As with the trapezoidal method, however, the end-correction term cannot be used if the slope approaches infinity. Furthermore, if the function has zero slope at the upper and lower limits, then the end correction term is zero.

In the next section we will consider a somewhat more complicated technique for numerical integration. We will implement this technique in a BASIC program and then run the program on the same three functions we examined in the section on Simpson's method. This will provide an opportunity for comparison of the two methods.

THE ROMBERG METHOD

The Simpson method, which uses a set of second-order equations, is an improvement over the trapezoidal method, in which first-order equations are used. Accordingly, we might attempt to further improve our numerical integration method by replacing the original curve with a set of cubic, or even higher-order, polynomials. There is another approach, however, known as the Romberg integration method. With this technique, the area is calculated by the trapezoidal method, but the errors inherent in the trapezoidal method are accounted for by using an interpolation method.

We designate the usual sequence of the trapezoidal values by the notation t_{11}, t_{21}, t_{31}, etc. They are assigned to the first column of the two-dimensional matrix T. The first level of interpolated values is designated as t_{12}, t_{22}, t_{32}, etc. They are placed into column 2 of the T matrix. We can then interpolate between the interpolated values to produce a third column designated t_{13}, t_{23}, t_{33}, etc.

$$\begin{bmatrix} t_{11} & t_{12} & t_{13} & \cdots \\ t_{21} & t_{22} & t_{23} & \cdots \\ t_{31} & t_{32} & t_{33} & \cdots \\ \cdots & \cdots & \cdots & \cdots \end{bmatrix}$$

If we continue in this way, we will find that the interpolated values rapidly converge upon the correct integral. The advantage of this method is that the function only has to be evaluated for the entries in the first column, corresponding to the regular trapezoidal-rule values. The function does not have to be evaluated to obtain the entries for the other columns. Rather, each value is obtained from a combination of the entry directly to the left and the one just below the entry to the left. For example:

$$t_{12} = \frac{4t_{21} - t_{11}}{3}$$

$$t_{22} = \frac{4t_{31} - t_{21}}{3}$$

$$t_{13} = \frac{16t_{22} - t_{12}}{15}$$

The general algorithm is:

$$T_{ij} = \frac{4^{j-1} t_{i+1, j-1} - t_{i, j-1}}{4^{j-1} - 1}$$

BASIC PROGRAM: INTEGRATION BY THE ROMBERG METHOD

A program that can be used to perform numerical integration by the Romberg method is given in Figure 9.14. Type up the program and run it.

```
10    REM Integration by Romberg's rule, Jul 17, 81
11    REM
12    REM identifiers
14    REM      D7      DELTA      panel width
15    REM      F2      FX         function to integrate
16    REM      L7      LOWER      lower limit
17    REM      P5%     PIECES%    number of panels
18    REM      S3      SUM
19    REM      T1      TOL        tolerance
20    REM      U2      UPPER      upper limit
21    REM end of identifiers
22    REM
30    DEF FNF2(X) = 1 / X
40    DIM N9%(16), T(136)
50    REM
60    L7 = 1
70    U2 = 9
80    T1 = .00001
90    GOSUB 2000
100   PRINT
110   PRINT " area  = "; S3
120   GOTO 9999
2000  REM integration by Romberg rule
2010  P5% = 1
2020  N9%(1) = 1
2030  D7 = (U2 − L7) / P5%
2040  C = (FNF2(L7) + FNF2(U2)) / 2
2050  T(1) = D7 * C
2060  N% = 1
2070  N8% = 2
2080  S3 = C
2090      N% = N% + 1
2100      F3 = 4
```

Figure 9.14: The Romberg Method

```
2110      N9%(N%) = N8%
2120      P5% = P5% * 2
2130      L% = P5% - 1
2140      D7 = (U2 - L7) / P5%
2150      FOR I1% = 1 TO (L% + 1 ) / 2
2160         I% = I1% * 2 - 1
2170         X = L7 + I% * D7
2180         S3 = S3 + FNF2(X)
2190      NEXT I1%
2200      T(N8%) = D7 * S3
2210      PRINT USING "### ##.#####"; P5%; T(N8%),
2220      N7% = N9%(N% - 1)
2230      K% = N% - 1
2240      FOR M% = 1 TO K%
2250         J% = N8% + M%
2260         N6% = N9%(N% - 1) + M% - 1
2270         T(J%) = (F3 * T(J% - 1) - T(N6%)) / (F3 - 1)
2280         F3 = F3 * 4
2290      NEXT M%
2300      PRINT USING "    ### ##.#####"; J%; T(J%)
2310      IF (N% < = 4) THEN 2360
2320         IF (T(N8% + 1) = 0) THEN 2360
2330            IF (ABS(T(N7% + 1) - T(N8% + 1)) < = ABS(T(N8% + 1) * T1))
                 THEN 2380
2340            IF (ABS(T(N8% - 1) - T(J%)) < = ABS(T(J%) * T1)) THEN 2380
2350            IF (N% > 15) THEN 2400
2360      N8% = J% + 1
2370   GOTO 2090
2380   S3 = T(J%)
2390   RETURN : REM normal return
2400   PRINT " no convergence"
2410   RETURN
9999   END
```

Figure 9.14: The Romberg Method (cont.)

First Run of the Romberg Program

The regular trapezoidal values are printed at each step, as before. In addition, the interpolated values at the right side of the matrix are also shown. The results should look like Figure 9.15.

```
            2  3.02222      3  2.54815
            4  2.46349      6  2.25919
            8  2.27341     10  2.20472
           16  2.21733     15  2.19773
           32  2.20234     21  2.19724
           64  2.19851     28  2.19723

         area  =    2.19723
```

Figure 9.15: Integration by the Romberg Method

By comparing the values in Figure 9.15 to those in the previous tables, it can be seen that the Romberg method converges even more rapidly (in this case) than Simpson's method.

Second Run of the Romberg Method—An Exponential Function

Alter the Romberg method so that it calculates the integral:

$$\int_0^1 e^{-x^2}\,dx$$

Change the function F2 at line 30 so that it reads:

30 **DEF** FNF2(X) = EXP($-$X$*$X)

and change the limits in the first line in the main program to read:

60 L7 = 0

70 U2 = 1

Run the Romberg method with the new formula. The results should look

```
            2   0.731370      3   0.747180
            4   0.742984      6   0.746834
            8   0.745866     10   0.746824
           16   0.746585     15   0.746824

         area  =    0.746824
```

Figure 9.16: Integration of e^{-x^2} by Romberg's Method

like Figure 9.16. Notice that Simpson's rule and the Romberg method perform equally well with this formula.

Third Run of the Romberg Method—A Periodic Function

We will now return to the integral:

$$\int_{0}^{4\pi} \sin^2 x \, dx$$

If we attempt to evaluate this integral using the Romberg method, we will essentially obtain a value of zero for the first two approximations. However, if we are careful to use a relative tolerance rather than an absolute tolerance to determine convergence, we will find the correct solution after several more iterations.

Make a copy of the Romberg program. Then change the three places that were changed in the Simpson method solution of this problem. The statement:

25 P7 = 3.14159

is added, and function F2 is altered to read:

30 **DEF** FNF2(X) = SIN(X)^2

The limits are set to:

60 L7 = 0
70 U2 = 4 * P7

Run the program and compare the output to Figure 9.17.

```
      2  9.64326e-12     3  8.57179e-12
      4  8.08070e-12     6  7.49238e-12
      8  6.28319        10  9.07793
     16  6.28319        15  6.08755
     32  6.28319        21  6.28633

   area  =   6.28633
```

Figure 9.17: Integration of sin²x by the Romberg Method

Finally, in the next section, we will expand our Romberg integration program to deal with a special case—a function that approaches infinity at one of the limits to the area beneath the curve. To solve this problem we will develop a technique of adjusting the width of the panels as they approach the infinite limit until we have arrived at a sufficiently accurate value for the total area.

FUNCTIONS THAT BECOME INFINITE AT ONE LIMIT

For each of the above methods we have used a uniform panel width throughout the desired interval; but it should be apparent that convergence is more difficult (that is, more panels are required) when the function has a greater slope. Conversely, wider panels can be used when the slope is smaller. Furthermore, a function may be infinite at one limit even though the area for the interval is finite.

Consider, for example, the integral of the reciprocal of the square root of x over the range from zero to unity:

$$\int_0^1 \frac{dx}{\sqrt{x}}$$

The exact value of the integral is found to be 2.0 by direct integration.

If we attempt to solve this problem with one of the previous integration programs we immediately run into a problem. The value of $f(a)$ at the left edge is infinite, and so we must choose some other lower limit. If we choose a small lower limit such as 10^{-11}, convergence will take a very long time. But if we choose a larger value, such as 0.01, for the left limit, the result will be inaccurate.

BASIC PROGRAM:
ADJUSTABLE PANELS FOR AN INFINITE FUNCTION

One method of dealing with this situation is to start the integration at a value somewhat larger than zero. We will initially choose the left boundary to be a value of 0.1. The integral is then evaluated over the range 0.1 to 1.0. To test the reasonableness of this, we should then evaluate the next region to the left, from 0.01 to 0.1. If this area is also significant, then we must add its area to the area obtained for the first regions. We then take the next region from 0.001 to 0.01 and see how large it is. In this way, we can take regions closer and closer to zero, observing the additional area as we go. The program shown in Figure 9.18 uses this approach with the Romberg method.

Running the Program

Type up the program and run it. The program is similar to the previous one, and so it may be easier to alter a copy of Figure 9.14.

The **PRINT** statements at lines 2210 and 2300 have been disabled so that only the final value of each major area is displayed. The main program repeatedly calls the subroutine at line 2000 with limits that are closer and closer to zero. When the new area is less than the tolerance, the process is terminated. The results should look like Figure 9.19.

```
10     REM Romberg Integration, Jul 17, 81
11     REM
12     REM identifiers
14     REM      D7      DELTA      panel width
15     REM      F2      FX         function to integrate
16     REM      L7      LOWER      lower limit
17     REM      P5%     PIECES%    number of panels
18     REM      P6      SUMT       total sum
19     REM      S3      SUM
20     REM      T1      TOL        tolerance
21     REM      U2      UPPER      upper limit
22     REM end of identifiers
30     DEF FNF2(X) = 1 / SQR(X)
40     DIM N9%(16), T(136)
50     A$ = " ##.######     ##.#####     ##.#^^^^     ##.#^^^^,,
60     L7 = .1
70     U2 = 1
80     T1 = .00001
90     P6 = 0
100    PRINT "  new area      total area     lower";
110    PRINT "   upper      limits"
120    GOSUB 2000
130    U2 = L7
140    L7 = .1 * U2
150    P6 = P6 + S3
160    PRINT USING A$; S3, P6, L7, U2
170    IF (ABS(S3) > T1) THEN 120
180    GOTO 9999
2000   REM integration by Romberg's rule
2010   P5% = 1
2020   N9%(1) = 1
2030   D7 = (U2 − L7) / P5%
2040   C = (FNF2(L7) + FNF2(U2)) / 2
2050   T(1) = D7 * C
2060   N% = 1
2070   N8% = 2
```

Figure 9.18: Romberg Integration with Adjustable Panels

```
2080    S3 = C
2090    N% = N% + 1
2100    F3 = 4
2110    N9%(N%) = N8%
2120    P5% = P5% * 2
2130    L% = P5% − 1
2140    D7 = (U2 − L7) / P5%
2150    FOR II% = 1 TO (L% + 1 ) / 2
2160        I% = II% * 2 − 1
2170        X = L7 + I% * D7
2180        S3 = S3 + FNF2(X)
2190    NEXT II%
2200    T(N8%) = D7 * S3
2210    REM PRINT USING "### ##.#####"; P5%; T(N8%),
2220    N7% = N9%(N% −1)
2230    K% = N% − 1
2240    FOR M% = 1 TO K%
2250        J% = N8% + M%
2260        N6% = N9%(N% −1) + M% − 1
2270        T(J%) = (F3 * T(J% −1) − T(N6%)) / (F3 − 1)
2280        F3 = F3 * 4
2290    NEXT M%
2300    REM PRINT USING "    ### ##.#####"; J%; T(J%)
2310    IF (N% < = 4) THEN 2360
2320        IF (T(N8% +1) = 0) THEN 2360
2330            IF (ABS(T(N7% +1)−T(N8% +1)) < = ABS(T(N8% +1) * T1))
                THEN 2380
2340            IF (ABS(T(N8% −1)−T(J%)) < = ABS(T(J%) * T1)) THEN 2380
2350            IF (N% > 15) THEN 2400
2360        N8% = J% + 1
2370    GOTO 2090
2380    S3 = T(J%)
2390    RETURN : REM normal return
2400    PRINT " no convergence"
2410    RETURN
9999    END
```

Figure 9.18: Romberg Integration with Adjustable Panels (cont.)

new area	total area	lower	upper	limits
1.36754	1.36754	1.0E-02	1.0E-01	
0.432456	1.80000	1.0E-03	1.0E-02	
0.136754	1.93675	1.0E-04	1.0E-03	
0.043246	1.98000	1.0E-05	1.0E-04	
0.013675	1.99368	1.0E-06	1.0E-05	
0.004325	1.99800	1.0E-07	1.0E-06	
0.001368	1.99937	1.0E-08	1.0E-07	
0.000432	1.99980	1.0E-09	1.0E-08	
0.000137	1.99994	1.0E-10	1.0E-09	
0.000043	1.99998	1.0E-11	1.0E-10	
0.000014	1.99999	1.0E-12	1.0E-11	
0.000004	2.00000	1.0E-13	1.0E-12	

Figure 9.19: Integration of $1/\sqrt{x}$ Near Zero

If the intermediate **PRINT** statements in the subroutine are not removed, it will be seen that each of the above regions has been divided into 64 panels. It should be realized that the overall width of each region is one-tenth that of the region immediately to the right and that the final left limit is 10^{-13}. Thus, if the whole region from 10^{-13} to 1.0 were integrated as one unit, it would require over a million million uniformly spaced panels. Such a method would take an extremely long time to obtain the correct answer.

SUMMARY

In this chapter we have developed and compared three different numerical integration programs. The first two, the trapezoidal rule and the Simpson method implementations, utilize ''end correction'' for refinement of the final approximation of the area under the curve. The third, more sophisticated method, Romberg's integration, uses a matrix of progressive interpolations, and does not require error correction. We tried these methods on several different functions; in addition, we used a refined Romberg integration program to compute the area under the curve of an infinite function.

All of the above integrations except the last were performed on analytic functions with uniformly spaced panels. Other techniques are required with discrete data. One approach is to fit the data to a polynomial function, using one of the techniques we have discussed in earlier chapters, then integrate the resulting polynomial.

EXERCISES

9-1: *Find the value of the integral:*

$$\int_0^{10} x^3 e^{-x}\, dx$$

by the trapezoidal rule, Simpson's rule, and the Romberg method. Which method seems best?

Answer: 5.938; Romberg

9-2: *Find the value of the integral:*

$$\int_0^{10} \frac{x\, dx}{1 + e^x}$$

by the trapezoidal rule, Simpson's rule, and the Romberg method. Which method seems best?

Answer: 0.8220; Simpson

9-3: *Find the value of the integral:*

$$\int_0^{10} \frac{dx}{e^x + e^{-x}}$$

by the trapezoidal rule, Simpson's rule, and the Romberg method. Which method seems best?

Answer: 0.785; trapezoid

9-4: *Find the value of the integral:*

$$\int_0^{10} \frac{dx}{(1 + x^2)^2}$$

by the trapezoidal rule, Simpson's rule, and the Romberg method. Which method seems best?

Answer: 0.785; trapezoid

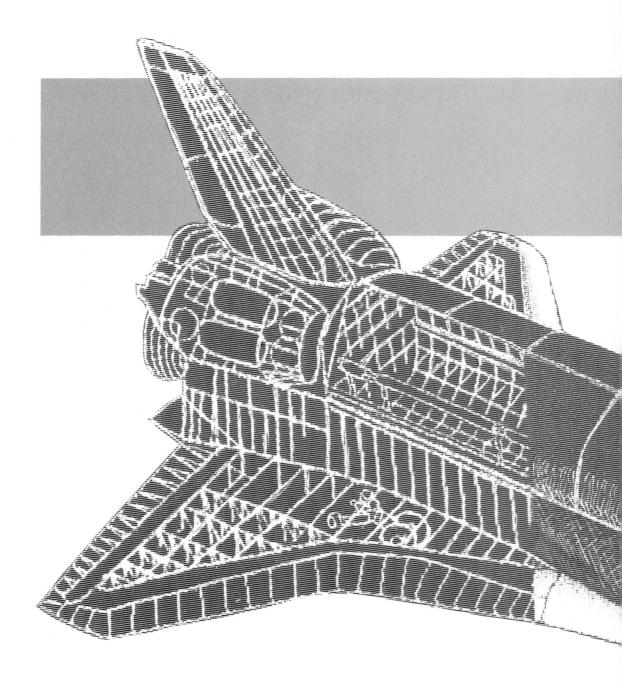

CHAPTER 10

Nonlinear Curve-Fitting Equations

INTRODUCTION

In Chapters 5 and 7 we developed computer programs for finding the coefficients to various curve-fitting equations. Since we chose approximating functions with linear coefficients, the resulting equations were linear, and therefore easily solved. Sometimes, however, it is necessary to choose approximating functions with nonlinear coefficients. In this case, the calculation of the coefficients may be more difficult.

In this chapter, we will consider two different techniques for nonlinear curve fitting. In one method, we will linearize the approximating function, and then find the solution to the linear form. Using this first method, we will find curve fits for two different sets of data. First, we will fit a linearized form of the so-called *rational function* to data representing the *Clausing factor*. Then, we will fit a linearized exponential function to data representing the diffusion of zinc in copper over a given temperature range.

Our second approach to nonlinear curve fitting will be more direct. We will see that there is no general technique for handling approximating functions that have nonlinear coefficients; however, we will investigate a specific method for exponential equations. This method involves eliminating one of the coefficients and then solving the resulting equations with Newton's method, which we studied in Chapter 8.

Let us begin, then, with our first example of the linearization method.

LINEARIZING THE RATIONAL FUNCTION

A commonly used, nonlinear approximating function is known as the *rational function*. This expression is formed from the ratio of two polynomials:

$$y = \frac{A_1 + A_3x + A_5x^2...}{1 + A_2x + A_4x^2 + A_6x^3...}$$

In this expression, x is the independent variable, y is the dependent variable and A_1, A_2, etc., are the coefficients as usual.

The rational function is nonlinear. However, it can be linearized by the following operations. Both sides of the equation are multiplied by the denominator polynomial, to give:

$$y(1 + A_2x + A_4x^2 + A_6x^3 ...) = A_1 + A_3x + A_5x^2 ...$$

The terms of the new equation can be rearranged to give:

$$y = A_1 - A_2xy + A_3x - A_4x^2y + A_5x^2 - A_6x^3y ...$$

Some of the terms on the right contain the dependent variable, y, as well as the independent variable, x. But remember, both x and y are arrays of known values. It is the unknown coefficients $A_1, A_2, ...A_n$ that we want. All of these coefficients are now linear, and so they can be determined by methods developed in Chapters 5 and 7.

BASIC PROGRAM:
THE CLAUSING FACTOR FITTED TO THE RATIONAL FUNCTION

A program for producing a least-squares fit to the linearized form of the rational function is given in Figure 10.1. The program can be derived from Figure 7.3 in Chapter 7.

The data in this program represent the *Clausing factor* as a function of length-to-radius (L/r) for cylindrical orifices. When molecules with a long mean free path effuse through a cylindrical orifice, some of the molecules strike the orifice walls and are returned in the opposite direction. The remaining molecules continue through to the other side of the orifice. The Clausing factor gives the fraction of those molecules entering one end of a cylindrical orifice that actually emerge from the other end. The Clausing factor ranges from zero to unity. Of course, the Clausing factor becomes smaller as the cylinder increases in length or decreases in radius. Thus, the ratio of length-to-radius, L/r, would appear in a formula for the direct calculation of the Clausing factor.

```
10    REM Least—squares fit to rational function, Jul 16, 1981
11    REM identifiers
13    REM    C1      COEF        solution vector
14    REM    C3      CORREL      correlation coefficient
15    REM    E2      SIGMA       vector of errors
16    REM    E5      SEE         std error of estimate
17    REM    M1%     MAX%        maximum length
18    REM    N1%     NROW%       number of rows
19    REM    N2%     NCOL%       number of columns
20    REM    R3      RESID       vector of residuals
21    REM    S7      SUMY        sum of y
22    REM    S8      SUMY2       sum y squared
23    REM    T6      SRS         sum residuals squared
24    REM    Y2      YCALC       calculated y
25    REM end of identifiers
30    REM
40    A$ = '' ### ##.# ##.#### ##.#### ##.####''
50    C$ = '' ##.#####  ##.####^^^''
60    M1% = 35
70    DIM Z(4), A(4,4), C1(4), Y(35), U(35,4)
80    DIM W(4,1), B(4,4), I2%(4,3), X(35), Y1(35)
90    DIM Y2(35), R3(35), E2(4)
100   REM
110   PRINT
120   PRINT '' Least—squares fit to the rational function''
130   REM
140   GOSUB 500 : REM get the data
150   REM sort the data
160   GOSUB 800 : REM set up the matrix
170   GOSUB 4000 : REM square up the matrix
180   GOSUB 5000 : REM Gauss—Jordan solution
190   GOSUB 1000 : REM print results
200   REM
210   GOTO 9999 : REM done
```

Figure 10.1: The Clausing Factor Fitted to the Ratio of Two Polynomials

```
500    REM get the data
510    N1% = 10
520    N2% = 4
530    FOR I% = 1 TO N1%
540       READ X(I%), Y1(I%)
550    NEXT I%
560    REM
570    DATA 0.1, 0.9524, 0.2, 0.9092
580    DATA 0.5, 0.8013, 1.0, 0.6720
590    DATA 1.2, 0.6322, 1.5, 0.5815
600    DATA 2.0, 0.5142, 3.0, 0.4201
610    DATA 4.0, 0.3566, 6.0, 0.2755
620    RETURN : REM from input routine
800    REM setup the data matrix
810    FOR I% = 1 TO N1%
820       X = X(I%)
830       Y = Y1(I%)
840       Y(I%) = Y
850       U(I%,1) = 1
860       U(I%,2) = − X * Y
870       U(I%,3) = X
880       U(I%,4) = −X * X * Y
890    NEXT I%
900    RETURN : REM from setting up data matrix
1000   REM Calculate residuals and print results
1010   S7 = 0
1020   S8 = 0
1030   T6 = 0
1040   FOR I% = 1 TO N1%
1050      X = X(I%)
1060      Y = Y(I%)
1070      Y2 = C1(3) − C1(4) * X * Y
1080      Y2 = C1(1) + (−C1(2) * Y + Y2) * X
```

Figure 10.1:
The Clausing Factor Fitted to the Ratio of Two Polynomials (cont.)

```
1090      R3(I%) = Y2 — Y(I%)
1100      Y2(I%) = Y2
1110      T6 = T6 + R3(I%) * R3(I%)
1120      S7 = S7 + Y(I%)
1130      S8 = S8 + Y(I%)*Y(I%)
1140   NEXT I%
1150   C3 = SQR(1—T6/(S8 — S7*S7/N1%))
1160   IF (N1% = N2%) THEN E5 = SQR(T6)
1170   IF (N1% <> N2%) THEN E5 = SQR(T6/(N1% — N2%))
1180   FOR J% = 1 TO N2%
1190      E2(J%) = E5 * SQR(ABS(B(J%,J%)))
1200   NEXT J%
1210   PRINT "   X      Y      Y Calc      Resid"
1220   FOR I% = 1 TO N1%
1230      PRINT USING A$; I%; X(I%), Y(I%), Y2(I%), R3(I%)
1240   NEXT I%
1250   PRINT
1260   PRINT "coefficients          errors"
1270   PRINT USING C$; C1(1), E2(1);
1280   PRINT " Constant term"
1290   FOR I% = 2 TO N2%
1300      PRINT USING C$; C1(I%), E2(I%)
1310   NEXT I%
1320   PRINT
1330   PRINT "Correlation  coefficient  is"; C3
1340   RETURN : REM from printout
4000   REM U and Y converted to A and Z
          (Continue with lines 4010—4140 of Figure 7.3.)
4150   RETURN : REM from square
5000   REM Gauss—Jordan matrix inversion and solution
          (Continue with lines 5010—6130 of Figure 4.6.)
6140   RETURN : REM from Gauss—Jordan subroutine
9999   END
```

Figure 10.1:
The Clausing Factor Fitted to the Ratio of Two Polynomials (cont.)

Running the Program

Type up the program and run it. The matrix is calculated for four terms, corresponding to a first-order numerator and a second-order denominator. Additional terms can be easily added to the approximating function. The values of M1%, N2%, and the corresponding dimension statement must be changed to reflect the actual number of terms. In addition, expressions such as:

$$U(I\%,5) = U(I\%,3) * X$$
$$U(I\%,6) = U(I\%,4) * X$$

must be added if additional terms are desired. The results shown in Figure 10.2 correspond to the equation:

$$y = \frac{1.0017 + 0.236x}{1 + 0.751x + 0.091x^2}$$

If the data are fitted with a regular polynomial function rather than the rational function, the resulting fit will be not as good (for the same number of coefficients in the approximating function).

I	X	Y	Y CALC	RESID
1	0.1	0.9524	0.9529	0.0005
2	0.2	0.9092	0.9090	-0.0002
3	0.5	0.8013	0.8006	-0.0007
4	1.0	0.6720	0.6719	-0.0001
5	1.2	0.6322	0.6324	0.0002
6	1.5	0.5815	0.5818	0.0003
7	2.0	0.5142	0.5146	0.0004
8	3.0	0.4201	0.4199	-0.0002
9	4.0	0.3566	0.3563	-0.0003
10	6.0	0.2755	0.2756	0.0001

```
coefficients      errors
1.00168           4.6031E-04   Constant  term
0.75153           1.3709E-02
0.23645           1.2200E-02
0.09099           5.1455E-03

Correlation  coefficient  is   0.999999
```

Figure 10.2: The Clausing Factor vs L/r Fitted to a Rational Function

In our next example of the linearization approach, we will examine an exponential equation. Later in this chapter we will fit the same equation using another, more direct approach.

LINEARIZING THE EXPONENTIAL EQUATION

One of the most common nonlinear equations has the form:

$$y = Ae^{Bx}$$

where x is the independent variable, y is the dependent variable, and A and B are the desired coefficients. This equation occurs widely throughout science and engineering because it is the solution to a first-order differential equation.

This equation can be linearized by taking the logarithm. The result is:

$$\ln y = \ln A + Bx$$

In this form, the dependent variable is $\ln y$ and the unknown coefficients are $\ln A$ and B. Since the new coefficients are linear, a least-squares fit can be obtained with any of the programs developed in Chapters 5 and 7.

BASIC PROGRAM: AN EXPONENTIAL CURVE FIT FOR THE DIFFUSION OF ZINC IN COPPER

Figure 10.3 gives a program for finding a least-squares fit to the linearized exponential equation. The program can be derived from the previous program in this chapter. Notice that the calculation of the standard error has been removed and the value of SRS is printed instead.

The data embedded in the input routine represent the diffusion of zinc in copper over the temperature range 600° to 900°C. The diffusion equation is:

$$D = D_0 e^{-Q/RT}$$

where D is the diffusion coefficient in cm sq/sec, D_0 is the diffusion constant in the same units, Q is the activation energy in cal/mole, R is the gas constant, and T is the temperature in degrees Kelvin. The independent variable, x, is the reciprocal of the temperature in degrees Kelvin. The dependent variable, y, is the logarithm of the the diffusion coefficient.

```
10   REM Linearized exponential, Jul 17, 1981
11   REM identifiers
13   REM     C1     COEF        solution vector
14   REM     C3     CORREL      correlation coefficient
15   REM     E2     SIGMA       vector of errors
16   REM     E5     SEE         std error of estimate
```

Figure 10.3: A Least-Squares Fit to the Linearized Exponential Equation

```
 17   REM      M1%    MAX%     maximum length
 18   REM      N1%    NROW%    number of rows
 19   REM      N2%    NCOL%    number of columns
 20   REM      R3     RESID    vector of residuals
 21   REM      T6     SRS      sum residuals squared
 22   REM      Y2     YCALC    calculated y
 23   REM end of identifiers
 30   G3 = 1.987 : REM gas constant
 40   A$ = " ### #### ##.####^^^ ##.####^^^"
 50   C$ = " ##.##### ##.####^^^"
 60   M1% = 35
 70   DIM Z(4), A(4,4), C1(4), Y(35), U(35,4)
 80   DIM W(4,1), B(4,4), I2%(4,3), X(35), Y1(35)
 90   DIM Y2(35), R3(35), E2(4), T(35), D(35)
100   REM
110   PRINT
120   PRINT " Least—squares fit to the diffusion equation"
130   REM
140   GOSUB 500 : REM get the data
150   REM sort the data
160   GOSUB 800 : REM set up the matrix
170   GOSUB 4000 : REM square up the matrix
180   GOSUB 5000 : REM Gauss—Jordan solution
190   GOSUB 1000 : REM Print results
200   REM
210   GOTO 9999 : REM done
500   REM get the data
510   N1% = 7
520   N2% = 2
530   FOR I% = 1 TO N1%
540      READ T(I%), D(I%)
542      X(I%) = 1 / (T(I%) + 273)
544      Y(I%) = LOG(D(I%))
550   NEXT I%
```

Figure 10.3:
A Least-Squares Fit to the Linearized Exponential Equation (cont.)

```
560    REM
570    DATA 600, 1.4E−12, 650, 5.5E−12
580    DATA 700, 1.8E−11, 750, 6.1E−11
590    DATA 800, 1.6E−10, 850, 4.4E−10
600    DATA 900, 1.2E−9
620    RETURN : REM from input routine
800    REM setup the data matrix
810    FOR I% = 1 TO N1%
850        U(I%,1) = 1
860        U(I%,2) = X(I%)
890    NEXT I%
900    RETURN : REM from setting up data matrix
1000   REM Calculate residuals and print results
1020   A1 = EXP(C1(1))
1030   T6 = 0
1040   FOR I% = 1 TO N1%
1050       Y = Y(I%)
1090       Y2(I%) = A1 * EXP(C1(2) * X(I%))
1100       IF (Y = 0) THEN R3(I%) = 1
1105       IF (Y <> 0) THEN R3(I%) = Y2(I%) / Y − 1
1110       T6 = T6 + R3(I%) * R3(I%)
1140   NEXT I%
1210   PRINT "    T    C      D      D Calc"
1220   FOR I% = 1 TO N1%
1230       PRINT USING A$; I%; T(I%), D(I%), Y2(I%)
1240   NEXT I%
1250   PRINT
1260   PRINT "Coefficients"
1270   PRINT C1(1);
1280   PRINT " Constant term"
1290   FOR I% = 2 TO N2%
1300       PRINT C1(I%)
1310   NEXT I%
1320   PRINT
```

Figure 10.3:
A Least-Squares Fit to the Linearized Exponential Equation (cont.)

```
1350   PRINT " D0 = "; EXP(C1(1)); " cm sq/sec"
1360   PRINT " Q = "; − G3 * C1(2) / 1000; " kcal/mole"
1364   PRINT
1365   PRINT "SRS = "; T6
1370   RETURN : REM from printout
4000   REM U and Y converted to A and Z
       (Continue with lines 4010 − 4140 of Figure 10.1.)
4150   RETURN : REM from square
5000   REM Gauss − Jordan matrix inversion and solution
       (Continue with lines 5010 − 6130 of Figure 4.6.)
6140   RETURN : REM from Gauss − Jordan subroutine
9999   END
```

Figure 10.3:
A Least-Squares Fit to the Linearized Exponential Equation (cont.)

Running the Program

Type up the program, run it, and compare the results with Figure 10.4. The diffusion constant can be calculated from the antilogarithm (the exponent) of the first coefficient with the following expression:

$$D0 = EXP(C1(1))$$

The activation energy is obtained from the second coefficient by using the expression:

$$Q = − G3 * C1(2)$$

where G3 is the gas constant.

In this example, the sum of residuals squared, *SRS*, is given rather than the usual standard error on the coefficients. The nonlinear transform of the approximating function makes these sigmas meaningless. The calculation of this form of *SRS* is discussed further in the next section, where we will develop our second approach to nonlinear curve fitting.

```
          T C        D            D Calc
     1    600    1.40000E-12    1.31283E-12
     2    650    5.50000E-12    5.43316E-12
     3    700    1.80000E-11    1.94311E-11
     4    750    6.10000E-11    6.13542E-11
     5    800    1.60000E-10    1.74041E-10
     6    850    4.40000E-10    4.49924E-10
     7    900    1.20000E-09    1.07266E-09

     Coefficients
     -1.13952          Constant  term
     -22889.4

     D0 =       0.32  cm sq/sec.
     Q  =      45.48  kcal/mole

     SRS  =    7
```

Figure 10.4: The Diffusion of Zinc in Copper (Linearized Fit)

DIRECT SOLUTION OF THE EXPONENTIAL EQUATION

In the previous section, the exponential equation:

$$y = Ae^{Bx}$$

was linearized by taking the logarithm:

$$\ln y = \ln A + Bx$$

A least-squares fit was then made to this linearized form of the equation. But the coefficients that produce the minimum sum of residuals squared to the linearized form will not, in general, produce the minimum *SRS* for the original, unlinearized equation. Thus we should consider a direct, least-squares solution to the nonlinear equation.

In this section, we will derive the curve-fitting equations with respect to the original nonlinearized exponential equation. If we approach the solution to the nonlinear equation the way we did for the linear equation, we will obtain the *SRS* for the approximating function. Then the derivative of the *SRS* with respect to each coefficient is set to zero. There is one equation for each unknown. Now, however, the resulting equations are nonlinear and therefore cannot generally be solved.

While there is no universal approach to the solution of nonlinear equations, there are several techniques that can be used for special forms. In the case of the exponential function, the solution is relatively easy. We will follow the usual curve-fit algorithm up to the point of taking the derivatives of the *SRS*. Then we will see a way to eliminate one of the coefficients from the equation, and we will solve the resulting function using Newton's method.

Calculating the SRS

The residuals for the equation:

$$y = Ae^{Bx}$$

could be defined as:

$$r = Ae^{Bx} - y$$

However, if the data are all measured to about the same relative degree of precision, independent of the magnitude, it will be more meaningful to use a relative residual:

$$r = (Ae^{Bx} - y) / y$$

or

$$r = (Ae^{Bx} / y) - 1$$

The residuals in the new form are squared, and then summed to form the *SRS*. The derivative of the resulting *SRS* is taken with respect to both *A* and *B*, and the resulting equations are set to zero. This approach is the same as before. There are two equations and two unknowns:

$$A\Sigma(e^{2Bx} / y^2) - \Sigma(e^{Bx} / y) = 0$$

and

$$A\Sigma(xe^{2Bx} / y^2) - \Sigma(xe^{Bx} / y) = 0$$

Now, however, the resulting equations are nonlinear, so they cannot be solved with the curve-fitting routine used in the previous section.

Eliminating Coefficient A

Fortunately, the coefficient *A* is linear in this case, and so it can be separated. The first of the two above equations can be rearranged to give:

$$A = \Sigma(e^{Bx} / y) / \Sigma(e^{2Bx} / y^2)$$

Then, this expression for *A* is substituted into the second equation to give:

$$\Sigma(e^{Bx} / y) \, \Sigma(xe^{2Bx} / y^2)$$

$$- \Sigma(e^{2Bx} / y^2) \, \Sigma(xe^{Bx} / y) = 0 = f(B)$$

Since we have eliminated A, this equation is only a function of B. In Chapter 8 we developed a program to solve nonlinear equations by Newton's method. The approach is applicable here.

Applying Newton's Method

We next take the derivative of the above equation with respect to B. The result is:

$$f'(B) = 2\Sigma(e^{Bx} / y) \, \Sigma(x^2 e^{2Bx} / y^2)$$

$$- \, \Sigma(xe^{2Bx} / y^2) \, \Sigma(xe^{Bx} / y)$$

$$- \, \Sigma(e^{2Bx} / y^2) \, \Sigma(x^2 e^{Bx} / y)$$

Before going any further, check to see that all of the terms in $f(B)$ and $f'(B)$ are homogeneous. The units of each term of $f(B)$ must correspond to:

$$\frac{x}{y^3}$$

and the units of $f'(B)$ must correspond to:

$$\frac{x^2}{y^3}$$

BASIC PROGRAM: A NONLINEARIZED EXPONENTIAL CURVE FIT

The program shown in Figure 10.5 can be used to find a nonlinear least-squares curve fit to the diffusion equation:

$$D = D_0 e^{-Q/RT}$$

This is the equation we considered in the previous section. Since we are using Newton's method, we need a first approximation for the value of B. We can obtain a good first value from the linear equation for B. But instead of performing the complete linearized fit, we can simply calculate the value of B from Equation 20 of Chapter 5. In this case, however, the value of y is replaced by $\ln y$:

$$B = \frac{\Sigma[x \ln (y)] - \Sigma(x) \, \Sigma[\ln (y) / n]}{\Sigma x^2 - (\Sigma x)^2 / n}$$

The first approximation for B is calculated at line 5130.

```
10    REM fit to nonlinearized exponential, Jul 17, 81
11    REM identifiers
13    REM      C1      COEF        solution vector
14    REM      E1%     ERMES%      error flag
15    REM      F       FX          function
16    REM      F1      DFX         function derivative
17    REM      H2      SMALL       small number
18    REM      M1%     MAX%        maximum length
19    REM      N1%     NROW%       number of rows
20    REM      N2%     NCOL%       number of columns
21    REM      R3      RESID       vector of residuals
22    REM      T1      TOL         tolerance
23    REM      T6      SRS         sum residuals squared
24    REM      Y2      YCALC       calculated y
25    REM end of identifiers
30    REM
32    F0% = 0
34    T0% = NOT F0%
35    T1 = .00001
37    H2 = 1E−15
40    A$ = '' ### ####  ##.####^^^ ##.####^^^''
50    C$ = '' ##.#####  ##.####^^^''
60    M1% = 35
65    G3 = 1.987 : REM gas constant
70    DIM Z(4), A(4,4), C1(4), Y(35)
80    DIM X(35), Y1(35), E9(35)
90    DIM Y2(35), R3(35), T(35), D(35)
100   REM
110   PRINT
120   PRINT ''Nonlinear least−squares fit''
130   REM
140   GOSUB 500 : REM get the data
150   REM sort the data
160   REM
```

Figure 10.5: A Least-Squares Fit to the Nonlinearized Exponential Function

```
170    GOSUB 5000 : REM nonlinear fit
180    REM
190    GOSUB 1000 : REM Print results
200    GOTO  9999 : REM done
210    REM
500    REM get the data
510    N1% = 7
520    N2% = 2
530    FOR I% = 1 TO N1%
540      READ T(I%), D(I%)
542      X(I%) = 1 / (T(I%) + 273)
544      Y(I%) = D(I%)
550    NEXT I%
560    REM
570    DATA  600, 1.4E−12, 650, 5.5E−12
580    DATA  700, 1.8E−11, 750, 6.1E−11
590    DATA  800, 1.6E−10, 850, 4.4E−10
600    DATA  900, 1.2E−9
620    RETURN : REM from input routine
1000   REM Calculate residuals and print results
1020   REM
1030   T6 = 0
1040   FOR I% = 1 TO N1%
1050     Y8 = Y(I%)
1090     Y2(I%) = A * E9(I%)
1100     IF (Y8 = 0) THEN R3(I%) = 1
1105     IF (Y8 <> 0) THEN R3(I%) = Y2(I%) / Y8 − 1
1110     T6 = T6 + R3(I%) * R3(I%)
1140   NEXT I%
1210   PRINT "      T   C      D       D Calc"
1220   FOR I% = 1 TO N1%
1230     PRINT USING A$; I%; T(I%), D(I%), Y2(I%)
1240   NEXT I%
1250   PRINT
```

Figure 10.5:
A Least-Squares Fit to the Nonlinearized Exponential Function (cont.)

```
1260    PRINT "Coefficients"
1270    PRINT C1(1);
1280    PRINT " Constant term"
1290    FOR I% = 2 TO N2%
1300       PRINT C1(I%)
1310    NEXT I%
1320    PRINT
1350    PRINT " D0 = "; A; " cm sq/sec"
1360    PRINT " Q = "; − G3 * C1(2) / 1000; " kcal/mole"
1364    PRINT
1365    PRINT "SRS = "; T6
1370    RETURN : REM from printout
5000    REM nonlinear fit to the exponential equation
5010    S3 = 0
5020    S7 = 0
5030    S9 = 0
5040    S4 = 0
5050    FOR I% = 1 TO N1%
5060       X8 = X(I%)
5070       Y8 = LOG(Y(I%))
5080       S3 = S3 + X8
5090       S7 = S7 + Y8
5100       S9 = S9 + X8 * Y8
5110       S4 = S4 + X8 * X8
5120    NEXT I%
5130    B = (S9 − S3 * S7/N1%)/(S4 − S3 * S3/N1%)
5140    GOSUB 8000 : REM Newton's method
5150    C1(1) = A
5160    C1(2) = B
5170    RETURN
8000    REM start of Newton's method
8005    X = B
8010    E1% = F0%
```

Figure 10.5:
A Least-Squares Fit to the Nonlinearized Exponential Function (cont.)

```
8020    FOR I% = 1 TO M1%
8030       X1 = X
8040       GOSUB 8400
8050       IF (ABS(F1) > H2) THEN 8090
8060       PRINT "ERROR—slope is zero"
8070       E1% = T0%
8080       GOTO 8160
8090       DX = F / F1
8100       X = X1 − DX
8110       REM PRINT "X = "; X1; ", fx = "; F; ", dfx = "; F1
8120       IF (ABS(DX) < = ABS(T1 * X)) THEN 8160
8130    NEXT I%
8140    PRINT "ERROR—no convergence in "; M1%; " tries"
8150    E1% = T0%
8160    RETURN : REM from Newton's method
8400    REM calculate function and its derivative
8405    B = X
8410    W1 = 0
8420    W2 = 0
8430    W3 = 0
8440    W4 = 0
8450    W5 = 0
8460    W6 = 0
8470    FOR J% = 1 TO N1%
8480       X8 = X(J%)
8490       Y8 = Y(J%)
8500       X2 = X8 * X8
8510       Y7 = Y8 * Y8
8520       W7 = EXP(B * X8)
8530       E9(J%) = W7
8540       W8 = W7 * W7
8550       W1 = W1 + X8 * W8 / Y7
8560       W2 = W2 + W7 / Y8
```

Figure 10.5:
A Least-Squares Fit to the Nonlinearized Exponential Function (cont.)

```
8570      W3 = W3 + X8 * W7 / Y8
8580      W4 = W4 + W8 / Y7
8590      W5 = W5 + 2 * X2 * W8 / Y7
8600      W6 = W6 + X2 * W7 / Y8
8610    NEXT J%
8900    F = W1 * W2 − W3 * W4
8910    F1 = W2 * W5 − W1 * W3 − W4 * W6
8915    A = W2 / W4
8920    RETURN
9999    END
```

Figure 10.5:
A Least-Squares Fit to the Nonlinearized Exponential Function (cont.)

Running the Program

Type up the program shown in the listing. The Newton's method routine is taken from lines 8000-8160 of Figure 8.12. The print statement at line 8110 is disabled. Notice that the Newton's method routine is working on the scalar value of X in place of the slope B. This is not to be confused with the array X. Run the program and compare the results to Figure 10.6.

Now, compare the results of Figure 10.4 to Figure 10.6. The diffusion constant D_0 and the activation energy Q are about the same. However,

```
   I     T C       D              D Calc
   1     600    1.40000E-12     1.31051E-12
   2     650    5.50000E-12     5.41377E-12
   3     700    1.80000E-11     1.93304E-11
   4     750    6.10000E-11     6.09471E-11
   5     800    1.60000E-10     1.72657E-10
   6     850    4.40000E-10     4.45808E-10
   7     900    1.20000E-09     1.06167E-09

   Coefficients
   .308916          Constant term
  -22860.3

   DO =     0.31 cm sq/sec
   Q  =    45.42 kcal/mole

   SRS =   0.0295
```

Figure 10.6: The Diffusion of Zinc in Copper (Nonlinear Fit)

the sum of residuals squared is smaller for the nonlinearized fit. It should be pointed out, however, that if the linearized residuals:

$$r = \ln D_0 - \frac{Q}{RT}$$

are used in *SRS*, the linearized *SRS* will be smaller. Furthermore, both linearized and nonlinearized programs will give exactly the same results if the data precisely follow an exponential equation. Therefore, it might be wise to perform both the linearized and nonlinearized curve fit. If the results are very different, then an error in measuring or in recording the data should be suspected.

SUMMARY

We have seen examples of two approaches to nonlinear curve fitting. The forms of the equations we used in our examples were (1) the rational function, and (2) the exponential function. We used a linearization approach and a direct approach. It is important to keep in mind that neither of the approaches we have studied represents a *general* method for nonlinear curve fitting; rather, we have examined specific techniques that can be used on specific curve-fitting equations.

EXERCISES

10-1: *The band-gap energy, E_g, of an intrinsic semiconductor can be determined from the expression:*

$$1/\rho = \sigma = Ae^{-(E_g/2KT)}$$

where ρ is the electrical resistivity, σ is the electrical conductivity, K is the Boltzmann constant, and T is the temperature. The following experimental data were obtained from an intrinsic thermistor:

T C	R ohms
0	1380
4	1200
10	880
18	660
25	488
38	304
43	248
54	170
62	139
77	83

Linearize the equation by taking the logarithm. Then perform a least-squares fit using the experimental data. Since we are only interested in the slope of the resulting fit, we do not have to convert resistance to resistivity. (Of course the temperature data must be converted to Kelvin.) Use a Boltzmann's constant of $8.61E-5$ to obtain the band gap energy in electron volts.

Answer: 0.6 ev.

10-2: *The time-temperature relationship for the crystallization of a material follows the expression:*

$$1/t = Ae^{-Q/RT}$$

where t is the time, Q is the activation energy, R is the gas constant and T is the temperature. Find the activation energy and coefficient A from the given data by performing a least-squares fit on the linearized form of the equation:

$$\ln t = -\ln A + Q/RT$$

Be sure to convert the temperature to Kelvin. For the gas constant use
$R = 1.987$ *cal/deg mole.*

T C	t min
350	49
360	40
370	34
380	28
390	24
400	20
410	17
420	14
430	12

Answer: $Q = 15.2$ Kcal/mole, $A = 4470$/min.

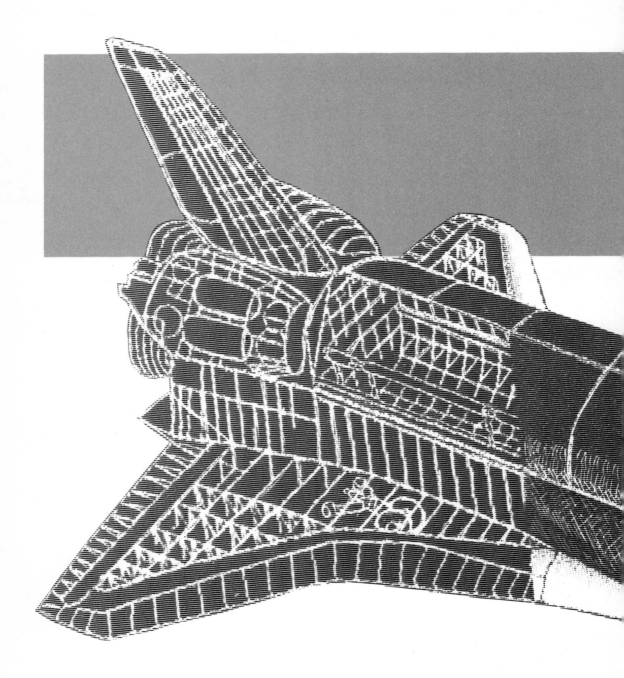

Chapter **11**

ADVANCED APPLICATIONS:

The Normal Curve, the Gaussian Error Function, the Gamma Function, and the Bessel Function

INTRODUCTION

This chapter takes up several advanced topics in programming for mathematical applications. The programs use a number of tools we have developed in this book. We will study three functions, all of which have important applications in mathematics, physics, and engineering: the Gaussian error function, the Gamma function, and the Bessel functions. For evaluating the Gaussian error function we will write two programs; the first uses Simpson's rule for numerical integration, and the second uses an infinite series expansion. In our discussion of the Gaussian error function we will again take up the topic of diffusion, which we studied in Chapter 10 in the context of nonlinear curve fitting. We will then examine the Gamma function. Following a study of the special properties of the function, we will develop a program to evaluate it. We will subsequently use this program in the last topic of the chapter—an investigation of numerical solutions to the Bessel equation. We will consider Bessel functions of the first and second kind.

The precision and range of a BASIC implementation will be significant issues in running the programs of this chapter. Thus, it will be important to recall the results of the evaluation programs we developed and ran in Chapter 1. Let us begin by reviewing the concepts of the distribution functions.

THE NORMAL AND CUMULATIVE DISTRIBUTION FUNCTIONS

We saw in Chapter 2 that random errors, introduced during experimental measurement, cause a sequence of observed values to be
dispersed about the mean. A frequency plot of the resulting data shows
a bell-shaped curve. This shape is described as a standard normal
distribution, or a probability density function, as in Figure 11.1.

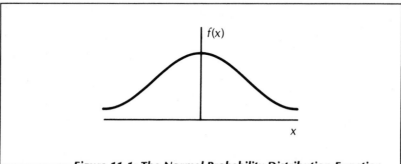

Figure 11.1: The Normal Probability Distribution Function

The normal distribution is defined by the equation:

$$f(x) = \frac{e^{\frac{-x^2}{2}}}{\sqrt{2\pi}} \tag{1}$$

This function has a peak, or mean value, at $x = 0$, and ranges from
minus infinity to plus infinity. The entire area under the probability-
density curve (above the x-axis) is normalized to a value of unity. That is,

$$\int_{-\infty}^{\infty} f(x)dx = 1 \tag{2}$$

From the symmetry of the curve, it can be seen that the area from $x = 0$
to infinity, that is, the right half of the curve, is equal to one-half the total
area.

The area from minus infinity to b is called the *cumulative distribution
function F(x)*.

$$F(x) = \int_{-\infty}^{b} f(x)dx \tag{3}$$

This integral cannot be solved in closed form, but it is tabulated in handbooks. Sometimes the integral:

$$G(x) = \int_0^b f(x)dx \qquad (4)$$

is given instead. Because the curve is normalized, either integral can readily be obtained from the other. The relationship is:

$$F(x) = G(x) + 0.5, \qquad x >= 0$$

In the next section we will consider some numerical methods for obtaining the area $G(x)$. But first, let us further explore the normal curve.

The Standard Deviation

The area under the normal curve of the measured values is related to the standard deviation. A small standard deviation corresponds to a close grouping of the measured values about the mean. Conversely, a large standard deviation corresponds to values that are spread further from the mean. Two normal distributions are shown in Figure 11.2. Both have the same mean value, but they have different standard deviations. The curve with the smaller standard deviation has the higher peak at zero. However, both have the same area underneath the curve.

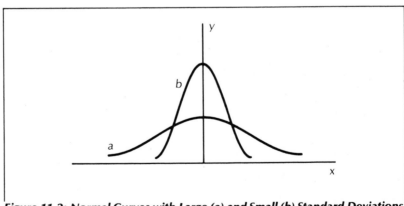

Figure 11.2: Normal Curves with Large (a) and Small (b) Standard Deviations

Figure 11.3 shows a plot of the distribution function described by Equation 3. This function has a value of 0.5 at $x = 0$ and rises asymptotically to a value of unity as x approaches infinity.

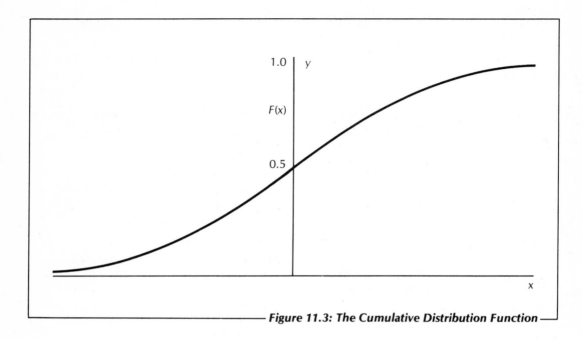

Figure 11.3: The Cumulative Distribution Function

The function $G(x)$ can be used to find the relationship between the standard deviation and the corresponding fraction of a particular sample. For example, $G(1)$ has a value of 0.34. Thus, 34% of the population lies in the range of zero to one standard deviation. Twice that value, 68%, corresponds to a range of one sigma on both sides of the mean. The distribution function can be readily obtained from the Gaussian error function, which we will discuss in the next section.

THE GAUSSIAN ERROR FUNCTION

Before formulating the Gaussian error function and implementing a program to evaluate it, we will digress slightly to a familiar application area—*diffusion*. The equation describing the diffusion of one kind of atom into another will finally lead us back to our Gaussian error function.

Diffusion in a One-Dimensional Slab

Diffusion is the net flow of atoms, electrons or heat from a region where it is more concentrated to a region where it is less concentrated. The resulting flux, J, can be described by *Fick's first law*:

$$J = -D \frac{dC}{dx}$$

where C is the concentration, and x is the distance. The *diffusion co-efficient* or *diffusivity*, D, is the same quantity we used to describe the diffusion of zinc in copper in Chapter 10. The minus sign reflects the fact that the flux occurs in a direction that is opposite to the concentration gradient. If we are concerned with the flow of atoms, then J is expressed in units of atoms per unit area-sec. The concentration, C, is given in units of atoms per unit volume.

Fick's first law can also be used to describe the flow of electrons in a conductor. We then write:

$$J = \sigma \frac{dV}{dx}$$

In this case, J is the current density, V is the voltage, and x is the distance. Sigma is the electrical conductivity.

In a similar way, the flow of heat can be expressed as:

$$J = k \frac{dT}{dx}$$

where k is the thermal conductivity in units of energy/area-sec, and T is the temperature.

Fick's second law:

$$\frac{\partial C}{\partial t} = D \frac{\partial^2 C}{\partial x^2}$$

can be used to describe the concentration as a function of position and time. There are many possible solutions to Fick's second law, depending on the boundary conditions. For example, steady-state conditions occur when the concentration no longer changes with time. Since

$$\frac{\partial C}{\partial t} = 0$$

then

$$\frac{D \partial^2 C}{\partial x^2} = 0$$

This implies that a concentration gradient, $\partial C/\partial x$ is linear.

Another useful solution to Fick's second law describes the diffusion of one kind of atom into another. The surface concentration of the diffusing species is kept constant for all time. The other species is in the shape of a one-dimensional, semi-infinite slab. In this case, the solution to Fick's law is:

$$\frac{C_x - C_0}{C_s - C_0} = 1 - \text{erf}(y) = 1 - \text{erf}\left(\frac{x}{\sqrt{2}\,Dt}\right)$$

In this expression, C_x is the concentration of the diffusing species at time t and a distance x from the surface. C_s is the surface concentration that is constant for all time, and C_0 is the initial, uniform concentration for all x at time equal to zero. D is the diffusion coefficient and y has the value:

$$\frac{x}{2\sqrt{Dt}}$$

If the initial concentration, C_0, is zero, then the equation reduces to:

$$\frac{C_x}{C_s} = 1 - \text{erf}(y) = 1 - \text{erf}\left(\frac{x}{2\sqrt{Dt}}\right)$$

The quantity erf is the Gaussian error function. It is defined as:

$$\text{erf}(y) = \frac{2}{\sqrt{\pi}} \int_0^y e^{-t^2} dt \tag{5}$$

The functions $F(x)$, given in Equation 3, and $G(x)$, given in Equation 4, can be obtained from the error function by the relationship:

$$F(x) = \frac{1 + \text{erf}\left(\frac{x}{\sqrt{2}}\right)}{2}$$

$$G(x) = \frac{\text{erf}\left(\frac{x}{\sqrt{2}}\right)}{2}$$

For example, the range represented by two standard deviations on either side of the mean can be found by the error function:

$$2G(2) = \text{erf}\left(\frac{2}{\sqrt{2}}\right) = 95\%$$

BASIC PROGRAM: EVALUATING THE GAUSSIAN ERROR FUNCTION USING SIMPSON'S RULE

The error function cannot be solved in closed form. However, it is possible to obtain particular solutions. A straightforward approach is to use a numerical integration technique such as Simpson's rule. Figure

11.4 gives a program for finding the error function in this way. It can be derived from Figure 9.8. The **PRINT** statements at lines 2080 and 2200 have been disabled so that intermediate results are not displayed.

```
10    REM Error function by Simpson's rule, Apr 21, 81
11    REM
12    REM identifiers
13    REM
14    REM        D7      DELTA       panel width
15    REM        E6      ESUM        end sum
16    REM        E8      EVSUM       even sum
17    REM        E9      ERF         error function
18    REM        F2      FX          function to integrate
19    REM        H3      SUM1
20    REM        L7      LOWER       lower limit
21    REM        O2      ODSUM       odd sum
22    REM        P5%     PIECES%     number of panels
23    REM        P7      TWOPI       2/sqr(pi)
24    REM        S3      SUM
25    REM        T1      TOL         tolerance
26    REM        U2      UPPER       upper limit
27    REM end of identifiers
28    REM
30    DEF FNF2(X) = EXP(−X∗X)
40    P7 = 2 / SQR(3.14159)
50    T1 = .00001
60    L7 = 0
70    INPUT '' Erf''; U2
80    IF (U2 < 0) THEN 9999
90    IF (U2 > 0) THEN 120
100       E9 = 0
110       GOTO 140
120    GOSUB 2000
130    E9 = P7 ∗ S3
140    PRINT
```

Figure 11.4: The Gaussian Error Function by Simpson's Rule

```
150   PRINT " Erf of "; U2; " is "; E9
160   GOTO 70
2000  REM integration by Simpson's rule
2010  P5% = 2
2020  D7 = (U2 − L7) / P5%
2030  O2 = FNF2(L7 + D7)
2040  E8 = 0
2050  E6 = FNF2(L7) + FNF2(U2)
2060  REM
2070  S3 = (E6 + 4 * O2) * D7 / 3
2080  REM PRINT USING " ##### #.#####"; P5%, S3
2090  P5% = P5% * 2
2100  H3 = S3
2110  D7 = (U2 − L7) / P5%
2120  E8 = E8 + O2
2130  O2 = 0
2140  FOR I% = 1 TO P5% / 2
2150    X = L7 + D7 * (2 * I% − 1)
2160    O2 = O2 + FNF2(X)
2170  NEXT I%
2180  S3 = (E6 + 4 * O2 + 2 * E8) * D7 / 3
2190  REM
2200  REM PRINT USING " ##### #.#####"; P5%, S3
2210  IF (ABS(S3 − H3) > ABS(T1 * S3)) THEN 2090
2220  RETURN
9999  END
```

Figure 11.4: The Gaussian Error Function by Simpson's Rule (cont.)

Running the Program

The program repeatedly cycles, asking the user for input. The error function is calculated by the Simpson method, then the argument and the corresponding error function are printed. The program can be terminated by entering a negative value.

There are several disadvantages to calculating the error function by this method. The execution time increases and the accuracy decreases as the argument increases. For a six- or seven-digit, floating-point package, the useful range of arguments is from zero to 3.

BASIC PROGRAM: EVALUATING THE GAUSSIAN ERROR FUNCTION USING AN INFINITE SERIES EXPANSION

Another way to evaluate the error function is to substitute an infinite series. The new expression is then integrated term by term to produce another infinite series. The result is:

$$\text{erf}(y) = \frac{2}{\sqrt{\pi}} \ e^{-y^2} \sum_{n=0}^{\infty} \frac{2^n y^{2n+1}}{1 \cdot 3 \cdot \ldots \cdot (2n+1)}$$

Figure 11.5 gives a program for evaluating the error function in this way. Type up the program and run it. The user is asked to input an argument to the error function. Then the argument and the resulting function are printed.

The infinite series is evaluated by the subroutine starting at line 2500. Each new term is added to the sum. If a particular term does not change the sum by more than the value of T1 (the tolerance), then the routine is terminated and the current value is returned.

```
10    REM Gaussian error function, Apr 21, 81
11    REM
12    REM identifiers
14    REM      E9      ERF          error function
15    REM      H3      SUM1
16    REM      P8      SQRTPI       sqr(pi)
17    REM      S3      SUM
18    REM      T1      TOL          tolerance
19    REM      T4      TERM
20    REM end of identifiers
21    REM
30    INPUT " Erf";  X
40    IF (X < 0) THEN 9999
50    GOSUB 2500
60    PRINT
70    PRINT " Erf of ";  X;  " is ";  E9
```

Figure 11.5: An Infinite Series Expansion for the Gaussian Error Function

```
  80    GOTO 30
2500    REM Gaussian error function by infinite series
2510    P8 = SQR(3.14159) : REM put in main program
2520    T1 = 1E−05
2530    IF (X < = 0) THEN 2670
2540    IF (X > 4) THEN 2690
2550    X2 = X * X
2560    S3 = X
2570    T4 = X
2580    I% = 0
2590    REM begin loop
2600        I% = I% + 1
2610        H3 = S3
2620        T4 = 2 * T4 * X2 / (1 + 2 * I%)
2630        S3 = T4 + H3
2640    IF (T4 > T1 * S3) THEN 2600
2650    E9 = 2 * S3 * EXP(−X2) / P8
2660    RETURN
2670    E9 = 0
2680    RETURN
2690    E9 = 1
2700    RETURN
9999    END
```

Figure 11.5:
An Infinite Series Expansion for the Gaussian Error Function (cont.)

Running the Program

The user must enter an argument that is equal to or larger than zero; the resulting function has a range from zero to unity. The result is zero if the argument is zero, and it approaches unity for arguments above 4. On the other hand, the function is approximately equal to its argument in the range of zero to 0.6. The error function has the same relative shape as the cumulative distribution function given in Figure 11.3. Selected values of the error function are given in Figure 11.6. The program will continually cycle, giving new values for the error function until a negative argument is entered.

y	erf(y)
0.0	0.0
0.1	0.1125
0.2	0.2227
0.3	0.3286
0.4	0.4284
0.5	0.5205
0.7	0.6778
1.0	0.8427
2.0	0.9953

Figure 11.6: The Gaussian Error Function

Suppose that zinc is to be diffused into a bar of copper at 900°C. We can use the error function to determine the concentration of copper as a function of time and the distance from the surface. The diffusion coefficient can be calculated from the diffusion constant and the activation energy. These latter quantities were found from the nonlinear curve fit in Chapter 10:

$$D = 0.31\ e^{\frac{-45,420}{1.987(900+273)}}$$

If the distance is chosen to be 0.01 cm and the time is taken as 13 hours, then the argument y in the error function becomes 0.7. This corresponds to an error function of 0.67 and an *error function complement* of 0.33. This means that the concentration of zinc at a depth of 0.01 cm will be 33% of the surface concentration after 13 hours.

In the next section we will see that round-off errors make it difficult to find a direct solution of the error function complement. We will examine a program that uses two functions to handle both small and large arguments.

THE COMPLEMENT OF THE ERROR FUNCTION

For the above solution to the diffusion equation, we are actually interested in the complement of the error function rather than the error function itself. The complement is defined as:

$$\text{erfc}(y) = \frac{2}{\sqrt{\pi}} \int_y^\infty e^{-t^2} dt \qquad (6)$$

The complement is obtained from the relationship:

$$\text{erfc}(y) = 1 - \text{erf}(y)$$

But if the complement of the error function is always calculated in this way, there will be large round-off errors for arguments above 3. As the error function approaches unity, the complement approaches zero. Ultimately, all significant figures are lost. Furthermore, the computation time increases as the argument increases.

We cannot integrate Equation 6 by using the trapezoidal rule or Simpson's method, either. A problem occurs in selecting the upper limit for the integral. A value larger than 8 will produce a floating-point underflow because e^{-64} is so small. Yet the area from this point to infinity is significant and cannot be ignored.

BASIC PROGRAM:
EVALUATING THE COMPLEMENT OF THE ERROR FUNCTION

One solution to this problem is to utilize two separate routines, one for small arguments and the other for larger arguments. The program given in Figure 11.7 uses this approach. The infinite-series expansion of erf, given in Figure 11.5, is used for smaller arguments. In addition, there is a second subroutine, beginning at line 2680, for larger arguments: the complement of the error function is calculated by means of an asymptotic expansion. This algorithm becomes more accurate as the argument increases. The equation, expressed as a continued fraction, rather than the usual infinite series, is:

$$\text{erfc}(y) = \frac{1/[1 + v/\{1 + 2v/[1 + 3v/(1 + \ldots)]\}]}{\sqrt{\pi}\ ye^{y^2}}$$

where

$$v = \frac{1}{2y^2}$$

10	**REM** Gaussian error function and complement, May 19, 81		
12	**REM** identifiers		
14	**REM**	C9	ERFC	complement
15	**REM**	E9	ERF	error function
16	**REM**	H3	SUM1	
17	**REM**	P8	SQRTPI	sqr(pi)

Figure 11.7: The Error Function and its Complement

```
  18   REM      S3      SUM
  19   REM      T1      TOL            tolerance
  20   REM      T4      TERM
  21   REM end of identifiers
  22   REM
  30   INPUT " Erf"; X
  40   IF (X < 0) THEN 9999
  50   GOSUB 2500
  60   PRINT
  70   PRINT " X = "; X; ", Erf = "; E9; ", Erfc = "; C9
  80   GOTO 30
2500   REM Gaussian error function by infinite series
2510   P8 = SQR(3.14159) : REM put in main program
2520   T1 = 1E−05
2530   IF (X < = 0) THEN 2680
2540   X2 = X * X
2550   IF (X > 1.5) THEN 2710
2560   S3 = X
2570   T4 = X
2580   I% = 0
2590   REM begin loop
2600      I% = I% + 1
2610      H3 = S3
2620      T4 = 2 * T4 * X2 / (1 + 2 * I%)
2630      S3 = T4 + H3
2640   IF (T4 > T1 * S3) THEN 2600
2650   E9 = 2 * S3 * EXP(−X2) / P8
2660   C9 = 1 − E9
2670   RETURN
2680   E9 = 0
2690   C9 = 1
2700   RETURN
2710   REM complementary error function
2720   T5 = 12
2730   V = 0.5 / X2
```

Figure 11.7: The Error Function and its Complement (cont.)

```
2740   U = 1 + V * (T5 + 1)
2750   FOR J% = T5 TO 1 STEP −1
2760     S3 = 1 + J% * V / U
2770     U = S3
2780   NEXT J%
2790   C9 = EXP(−X2) / (X * S3 * P8)
2800   E9 = 1 − C9
2810   RETURN
9999   END
```

Figure 11.7: The Error Function and its Complement (cont.)

Running the Program

Type up the new version and run it. The program will now print the error function and its complement. For arguments that are less than or equal to 1.5, the error function is calculated by the infinite series expansion given in Figure 11.5. The complement is then obtained by subtraction from unity. On the other hand, if the argument is larger than 1.5, the complement is calculated from the asymptotic expansion. The error function is determined by subtraction from unity.

Compare the error function from this version with the data in Figure 11.6. Then try the values given in Figure 11.8 to check the complement of the error function.

In the next section we will examine the special properties of the Gamma function and we will implement a program to evaluate it.

y	$erfc(y)$
1.5	$3.390E-2$
2.0	$4.678E-3$
2.5	$4.070E-4$
3.0	$2.209E-5$
3.5	$7.431E-7$
4.0	$1.542E-8$
4.5	$1.966E-10$

Figure 11.8: The Complement of the Error Function

THE GAMMA FUNCTION

A function that is related to the error function is known as the Gamma function. It is defined by the integral:

$$\Gamma(n) = \int_0^\infty x^{n-1} e^{-x} dx \tag{7}$$

The Gamma function is important because it is part of the solution to Bessel's equation. In addition, it can be used to calculate factorials because of the recursive relationship:

$$\Gamma(n+1) = n\Gamma(n) \tag{8}$$

Since $\Gamma(1) = 1$, we can see that

$$\Gamma(2) = \Gamma(1) = 1 \qquad = 1!$$
$$\Gamma(3) = 2\Gamma(2) = 1 \cdot 2 \quad = 2!$$
$$\Gamma(4) = 3\Gamma(3) = 1 \cdot 2 \cdot 3 = 3!$$
$$\cdots$$
$$\Gamma(n) = (n-1)\Gamma(n-1) = (n-1)!$$

Thus, the general formula for using the Gamma function to calculate factorials is:

$$\Gamma(n+1) = n\Gamma(n) = n!$$

Since the Gamma function is defined for all real arguments greater than zero, the "factorial" of noninteger arguments can be defined as well. In addition, we find that:

$$\Gamma(0.5) = -0.5! = \sqrt{\pi}$$

This will be useful in calculating Bessel functions. The Gamma function is also defined for noninteger negative numbers. Its value, however, is infinite for zero and negative integers. A plot of the Gamma function for real arguments is given in Figure 11.9.

The Gamma function can be calculated using a form of Stirling's approximation:

$$\Gamma(x) = \sqrt{\frac{2\pi}{x}} \, x^x e^y$$

where

$$y = \frac{1}{12x} - \frac{1}{360x^3} - x$$

Since this is an asymptotic series, the relative accuracy improves as the argument increases.

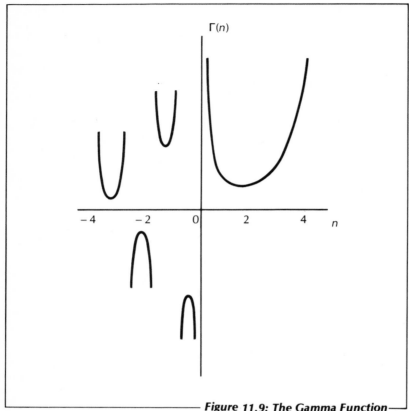

Figure 11.9: The Gamma Function

BASIC PROGRAM: EVALUATION OF THE GAMMA FUNCTION

The program given in Figure 11.11 uses Stirling's approximation to calculate the Gamma function. Arguments can be any real number greater than zero and less than about 32. The upper limit depends on the floating point arithmetic of your BASIC. Arguments can also be negative if they are nonintegral. However, the argument cannot be zero or a negative integer.

Positive arguments are incremented by 2, and the Gamma function of the new argument is calculated. The resulting Gamma function is then reduced to the corresponding original argument by using the algorithm:

$$\Gamma(x) = \frac{\Gamma(x+2)}{x(x+1)}$$

This conversion is not needed for larger arguments, but it insures that there will be at least six figures of precision for all values of x.

Negative arguments are incremented until they are positive. The Gamma function is then called with the new argument. This appears to be a recursive routine, but in fact it is not. Consequently, it can be implemented in standard BASIC, which does not allow recursive routines. The result is corrected to the original argument.

Running the Program

Type up the program and run it. Try the values given in Figure 11.10 and compare the values for Γ(x) with your results. The program cycles repeatedly until a value of zero is entered. Since:

$$x! = \Gamma(x + 1)$$

the program can be readily rewritten so that it will generate factorials rather than the Gamma function.

x	Γ(x)	
1	1	0!
2	1	1!
3	2	2!
4	6	3!
5	24	4!
6	120	5!
0.5	1.7725	$\sqrt{\pi}$
− 0.5	− 3.5449	− Γ(0.5) / 0.5
− 1.5	2.3633	− Γ(− 0.5) / 1.5

Figure 11.10: Selected Gamma Function Values

```
10   REM Gamma function, Apr 21, 81
12   REM identifiers
14   REM      G5      GAMMA
15   REM      H3      SUM1
16   REM      P7      PI
17   REM end of identifiers
20   REM
```

Figure 11.11: Evaluation of the Gamma Function

```
  30   INPUT " X ";  X
  40   IF (X = 0) THEN 9999
  50   GOSUB 9200
  60   PRINT
  70   PRINT " Gamma is ";  G5
  80   GOTO 30
9200   REM Gamma function
9210   P7 = 3.14159
9220   IF (X < 0) THEN 9270
9230   Y = X + 2
9240   G2 = SQR(2*P7/Y)*EXP(Y*LOG(Y) + (1-1/(30*Y*Y))/(12*Y)-Y)
9250   G5 = G2 / (X * (X + 1))
9260   RETURN
9270   REM negative argument
9280   K5% = 0
9290   G3 = X
9300   IF (X > = 0) THEN 9340
9310      K5% = K5% + 1
9320      X = X + 1
9330   GOTO 9300
9340   GOSUB 9200 : REM almost recursive
9350   FOR J% = 1 TO K5%
9360      X = X - 1
9370      G5 = G5 / X
9380   NEXT J%
9390   X = G3
9400   RETURN
9999   END
```

Figure 11.11: Evaluation of the Gamma Function (cont.)

In the final two sections of this chapter we will introduce the Bessel functions of the first and second kind. *Bessel's equation* has many mathematical and scientific applications; solutions to this differential equation can be found using the Bessel functions. We will examine two BASIC implementations of these functions.

BESSEL FUNCTIONS

Bessel's equation:

$$x^2y'' + xy' + (x^2 - n^2)y = 0$$

arises in the analysis of many different kinds of problems involving circular symmetry. In this equation, x is the independent variable, y is the dependent variable and n is a constant known as the order. This is a nonlinear differential equation that cannot be solved in closed form. One of the solutions to Bessel's equation is:

$$y = J_n(x)$$

where J is the n-order Bessel function of the first kind.

The Bessel functions have been extensively tabulated for particular values of x and n. However, they are difficult to use in this form. For values of x less than about 15, the J Bessel functions can be calculated from the infinite series:

$$J_n(x) = \sum_{k=0}^{n} \frac{(-1)^k}{k!\Gamma(n+k+1)} \left(\frac{x}{2}\right)^{n+2k} \tag{9}$$

On the other hand, the asymptotic expression:

$$J_n(x) = \sqrt{\frac{2}{\pi x}} \cos\left(x - \frac{\pi}{4} - \frac{n\pi}{2}\right) \tag{10}$$

can be used for larger values of x. The Bessel functions J_0 and J_1 are shown in Figure 11.12.

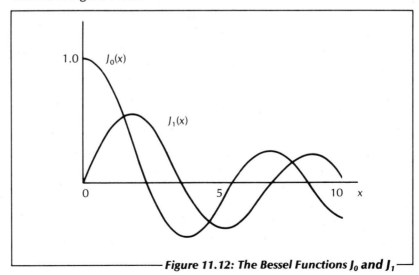

Figure 11.12: The Bessel Functions J_0 and J_1

BASIC PROGRAM: BESSEL FUNCTIONS OF THE FIRST KIND

The program shown in Figure 11.14 uses both Equations 9 and 10 to calculate the Bessel functions of the first kind. The order can be zero or a positive number. The argument can also be a noninteger negative number. The Gamma function from the previous section is incorporated. The infinite series is utilized for arguments less than 15, and the asymptotic expression is used for larger arguments. There may be inaccuracies in this transition region depending on the floating-point arithmetic of your BASIC.

Type up the program and run it. Try the values in Figure 11.13 for the order and argument, and compare the results. The program can be terminated by entering an order less than -25.

Order	Argument	Function
1	1	0.4401
0	1	0.7652
1	0.5	0.2423
0	0.5	0.9385
1	10	0.04347
0	10	-0.2459
0.25	1	0.7522
0.25	1.5	0.6192
0.5	1.5708 $(\pi/2)$	0.6366 $(2/\pi)$
-0.25	1	0.6694
-0.25	1.5	0.3180
-0.75	1.5	-0.2684

Figure 11.13: Selected Bessel Function Values

```
10    REM Bessel function, first kind, Apr 21, 81
12    REM identifiers
14    REM      D9    ORD        order
15    REM      G5    GAMMA
16    REM      J7    JBES       J Bessel function
17    REM      P7    PI
18    REM      S6    SUM
19    REM      T1    TOL        tolerance
```

Figure 11.14: Bessel Functions of the First Kind

```
20   REM      T3      NTERM      next term
21   REM      T4      TERM
22   REM end of identifiers
23   REM
30   INPUT " Order";  D9
40   IF (D9 < −25) THEN 9999
50   INPUT " X ";  X
60   GOSUB 8500
70   PRINT
80   PRINT " J Bessel is ";  J7
90   GOTO 30
8500 REM Bessel function of the first kind
8510 T1 = .00001
8520 P7 = 3.14159
8530 X2 = X * X
8540 IF ((X = 0) AND (D9 = 1)) THEN 8740
8550 IF (X > 15) THEN 8760
8560 IF (D9 <> 0) THEN 8590
8570 S6 = 1
8580 GOTO 8640
8590 X3 = X : REM save x
8600 X = D9 + 1 : REM Gamma needs x
8610 GOSUB 9200 : REM Gamma function
8620 X = X3 : REM restore x
8630 S6 = (X/2)^D9 / G5
8640 T3 = S6
8650 I% = 0
8660 IF (ABS(T3) < = ABS(S6*T1)) THEN 8720
8670 I% = I% + 1
8680 T4 = T3
8690 T3 = −T4 * X2 * .25/(I% * (D9 + I%))
8700 S6 = S6 + T3
8710 GOTO 8660
8720 J7 = S6
```

Figure 11.14: Bessel Functions of the First Kind (cont.)

```
8730   RETURN
8740   J7 = 0
8750   RETURN
8760   REM asymptotic expansion
8770   J7 = SQR(2/(P7 * X)) * COS(X − P7/4 − D9 * P7/2)
8780   RETURN
9200   REM Gamma function
9210   REM P7 = 3.14159
       (Continue with lines 9220 − 9390 of Figure 11.11.)
9400   RETURN
9999   END
```

Figure 11.14: Bessel Functions of the First Kind (cont.)

Since Bessel's equation is a second-order equation, two independent solutions are needed. When the order n is not an integer, then both solutions can be obtained from the J Bessel functions. The expression is:

$$y = AJ_n(x) + BJ_{-n}(x)$$

where A and B are constants to be determined from the boundary conditions.

BASIC PROGRAM: BESSEL FUNCTIONS OF THE SECOND KIND

If the order of Bessel's equation is an integer, then the solutions $J_n(X)$ and $J_{-n}(x)$ are linearly dependent. In this case, a second, independent solution can be obtained from the expression:

$$y = AJ_n(x) + BY_n(x)$$

where Y is a Bessel function of the second kind.

The program given in Figure 11.17 can be used to calculate Bessel functions of the second kind. While the order can be any real number, it is customary to use these functions only for integer orders. Two different algorithms are utilized. For arguments less than 12, the values of Y_0 and Y_1 are calculated from the expressions:

$$Y_0(x) = \frac{2}{\pi} \sum_{m=0}^{n} (-1)^m \left(\frac{x}{2}\right)^{2m} \frac{[\ln\left(\frac{x}{2}\right) + \gamma - h]}{(m!)^2}$$

and

$$Y_1(x) = -\frac{2}{\pi x} + \frac{2}{\pi} \sum_{m=1}^{n}(-1)^{m+1}\left(\frac{x}{2}\right)^{2m-1}\frac{\left[\ln\left(\frac{x}{2}\right)+\gamma-h+\frac{1}{2m}\right]}{(m!)(m-1)!}$$

where:

$$h = \sum_{r=1}^{m}\frac{1}{r} \quad \text{if} \quad m \geqslant 1$$

and γ is Euler's constant ($\gamma = 0.57721566$). Orders other than 0 and 1 are calculated from the formula:

$$Y_n(x) = \frac{2n}{x}Y_{n-1}(x) - Y_{n-2}(x)$$

For larger arguments, an asymptotic expansion similar to the one used for the J Bessel functions is employed:

$$Y_n(x) = \sqrt{\frac{2}{\pi x}}\sin\left(x - \frac{\pi}{4} - \frac{n\pi}{2}\right)$$

A plot of the functions Y_0 and Y_1 is given in Figure 11.15.

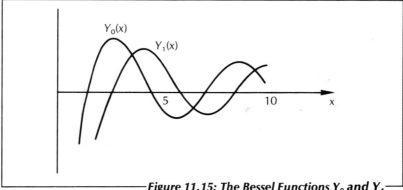

Figure 11.15: The Bessel Functions Y_0 and Y_1

Type up the program and try it out. The argument must be a positive number, since the function goes to minus infinity at zero. Some typical values are shown in Figure 11.16.

The program repeatedly cycles. It can be terminated by entering an order less than zero.

Argument	Y_0	Y_1
1	0.088	−0.781
2	0.510	−0.107
3	0.377	0.325
11	−0.169	0.164
15	0.206	0.021

Figure 11.16: Selected Values of the Bessel Function of the Second Kind

```
 10    REM Bessel function, second kind, May 23, 81
 12    REM identifiers
 14    REM      C5      EULER
 15    REM      D9      ORD          order
 16    REM      G5      GAMMA
 17    REM      H2      SMALL        small number
 18    REM      P7      PI
 19    REM      P8      PI2          2 / pi
 20    REM      S6      SUM
 21    REM      T4      TERM
 22    REM      Y8      YBES
 23    REM end of identifiers
 24    REM
 30    PRINT
 40    INPUT " Order";  D9
 50    IF (D9 < 0) THEN 9999
 60    INPUT " Arg";  X
 70    IF (X <= 0) THEN 60
 80    GOSUB 4400
 90    PRINT " Y Bessel is ";  Y8
100    GOTO 30
4400   REM Bessel function of the second kind
4410   H2 = 1E−08
4420   C5 = .577216
```

Figure 11.17: The Bessel Function of the Second Kind

```
4430    P7 = 3.14159
4440    P8 = .63662 : REM 2/pi
4450    IF (X > 12) THEN 4910
4460    Z5 = .5 * X
4470    X2 = Z5 * Z5
4480    T = LOG(Z5) + C5
4490    S6 = 0 : REM calculation of Y0
4500    T4 = T
4510    Y0 = T
4520    J% = 0
4530    IF (ABS(T4) < = H2) THEN 4600
4540      J% = J% + 1
4550      IF (J% <> 1) THEN S6 = S6 + 1.0/(J% −1)
4560      Z6 = T − S6
4570      T4 = −X2 * T4 / (J% * J%) * (1.0 − 1.0/(J% * Z6))
4580      Y0 = Y0 + T4
4590    GOTO 4530
4600    Y0 = P8 * Y0
4610    IF (D9 = 0) THEN 4870
4620    T4 = Z5 * (T − .5) : REM calculate Y1
4630    S6 = 0
4640    Y1 = T4
4650    J% = 1
4660    IF (ABS(T4) < = H2) THEN 4740
4670      J% = J% + 1
4680      S6 = S6 + 1.0/(J% −1)
4690      Z6 = T − S6
4700      T2 = −X2 * T4 / (J% * (J% −1))
4710      T4 = T2 * ((Z6 − .5/J%) / (Z6 + .5/(J% −1)))
4720      Y1 = Y1 + T4
4730    GOTO 4660
4740    Y1 = P8 * (Y1 − 1/X)
4750    IF (D9 = 1) THEN 4890
4760    REM find Yn by recursion with Y0 and Y1
4770    Z6 = 2 / X
```

Figure 11.17: The Bessel Function of the Second Kind (cont.)

```
4780   Z1 = Y0
4790   Z2 = Y1
4800   FOR J% = 2 TO INT(D9 + .01)
4810      Z3 = Z6 * (J% −1) * Z2 − Z1
4820      Z1 = Z2
4830      Z2 = Z3
4840   NEXT J%
4850   Y8 = Z3
4860   RETURN
4870   Y8 = Y0
4880   RETURN
4890   Y8 = Y1
4900   RETURN
4910   REM asymptotic expansion
4920   Y8 = SQR(2/(P7 * X)) * SIN(X − P7/4 − D9 * P7/2)
4930   RETURN
9999   END
```

Figure 11.17: The Bessel Function of the Second Kind (cont.)

SUMMARY

In Chapter 11 we have reviewed a number of the concepts and tools examined in this book. With the tools that we now have available to us, we have found that we can implement some rather advanced mathematical applications. We have seen BASIC programs that evaluate several variations of the Gaussian error function, the Gamma function, and the Bessel functions. In the process of developing these and other programs in this book, we have demonstrated how BASIC can be used for technical applications.

EXERCISES

11-1: *Make a copy of the program shown in Figure 11.5 and alter the new version so that it calculates the function G(x) described by Equation 4. Show that G(2) = 0.4772.*

11-2: *Copy Figure 11.10 and alter the new version so that it calculates factorials. Show that the factorial of 0.5 is 0.886.*

11-3: *Write a program to calculate combinations. Start with the program used in the previous problem. The number of ways that six things can be taken two at a time, 6C2, is calculated from the expression:*

$$\frac{6!}{2!(6-2)!}$$

The program should request two numbers: the total number of items and the number of items taken at one time. Show that 6C2 has a value of 15.

Reserved Words
and Functions

RESERVED WORDS

The following BASIC reserved words appear in **boldface** in this book:

AND	**DATA**	**DEF**	**DIM**	**ELSE**
END	**FOR**	**GOSUB**	**GOTO**	**IF**
INPUT	**LET**	**NEXT**	**NOT**	**ON**
OR	**PRINT**	**READ**	**REM**	**RESTORE**
RETURN	**STEP**	**STOP**	**THEN**	**TO**
USING				

These additional reserved words are commonly implemented in some BASICs:

DEFDBL	**DEFINT**	**DEFSTR**	**HEX$**	**LINE INPUT**
LPRINT	**MOD**	**OCT$**	**PEEK**	**POKE**
TROFF	**TRON**	**WEND**	**WIDTH**	**WHILE**

BUILT-IN FUNCTIONS

Name	Action
ABS	Absolute value
ASC	Convert ASCII character to integer equivalent
ATN	Arctangent
CHR$	Convert integer to ASCII equivalent
COS	Cosine
EOF*	True if end of disk file
EXP	Exponent of argument
FRE	Free bytes in memory
INSTR*	Find substring within a string
INT	Convert to integer
LEFT$	Left part of string
LEN	Length of string
LOG	Natural logarithm
MID$	Middle part of string
RIGHT$	Right part of string
RND	Random number
SGN	Sign of argument times unity
SIN	Sine
SPACE$*	Make a string of spaces
SPC	Print spaces
SQR	Square root
STRING$	Make a string from the given character
STR$	Convert real number to string
SWAP*	Interchange two values
TAB	Move to indicated print position
TAN	Tangent
VAL	Convert string to real number

---

*Microsoft BASIC

APPENDIX **B**

Summary of BASIC

THE BASIC CHARACTER SET

A full alphabet	A-Z, a-z
The digits	0-9
The special characters	$+ - * / = < > ( ) . , ; : ^ \$ \% ''$
A space or blank character	

VARIABLE NAMES

A BASIC variable name may be an upper-case letter or an upper-case letter followed by a digit. Many implementations allow any alphanumeric character in the second position. Some BASICs allow names to be as long as 31 characters.

Examples:

A, B, A1, B2	All BASICs
UP, BY	Many BASICs
JUMP, MAX	When long names are allowed

Variables default to the type REAL unless the name contains a suffix. A \$ suffix is used to denote a string variable. A % suffix is commonly used to designate an integer variable.

Examples:

UPPER\$, LOWER\$	String variables
LOW%, HIGH%	Integer variables

ARRAY VARIABLES

BASIC arrays may have one, two, or more dimensions. The convention for naming array variables is usually the same as it is for scalar variables. The array size and number of dimensions are declared in a **DIM** statement. Array elements are referenced by an integer index number that runs from zero to the maximum value declared in the dimension statement.

Examples:

DIM TEMP(20)	(A vector with 21 elements)
DIM A(8,8)	(A 9-by-9 matrix)
DIM LOWER$(20)	(An array of 21 strings)

CONSTANTS

BASIC uses numeric and string constants. Numeric constants can be written as integers or as real numbers. Integers are written as strings of digits. Some BASICs use a % suffix to denote an integer constant. String constants are embedded in quotation marks.

Examples:

$$15 \qquad -19253 \qquad +7 \qquad 0 \qquad 5\% \qquad \text{"Friday"}$$

Real numbers are written with a decimal point, a scale factor, or both:

$$379.1275 \qquad 3.791275E2 \qquad 3791275E-4$$

The E notation indicates multiplication by powers of 10. Thus, E2 signifies 100, and E–4 signifies .0001.

COMMENTS

Comments begin with the reserved word **REM** (for remark). The remainder of the line is ignored by BASIC.

Example:

REM Plot Y and Y2 as a function of X

OPERATIONS

*Arithmetic Operators*_____

^	exponentiation
+	addition
—	subtraction and negation
*	multiplication
/	division
\	integer division (Microsoft)

*Relational Operators*_____

These operations result in a value of zero (false) or not zero (true):

=	equality
<>	inequality
<	less than
>	greater than
< =	less than or equal to
> =	greater than or equal to

*Logical Operators*_____

These operations result in a value of zero (false) or not zero (true):

AND
OR
NOT
XOR

*Functional Operators*_____

The built-in functions are given in Appendix A. Additional functions can be written as needed. The letters FN are prefixed to the function name.

Example:

DEF FNMAGNITUDE(X,Y) = SQR(X * X + Y * Y)

ASSIGNMENT STATEMENTS

Scalar variables and individual array elements are assigned values with the form:

VARIABLE = EXPRESSION

where EXPRESSION is a constant, a variable or a combination of constants and variables resulting in a single value. EXPRESSION must match the type of VARIABLE. That is, a string value cannot be assigned to a numerical variable and vice versa. The relational operator (=) should be carefully distinguished from the assignment operator (=). Notice that the expression:

A = B = C

assigns the result of TRUE to the variable A if B and C are the same. The value of A becomes FALSE otherwise.

THE UNCONDITIONAL BRANCH

The **GOTO** statement is an unconditional branch. The orderly flow of statement execution is interrupted and the statement at the indicated line number is executed next.

Example:

GOTO 9999

Multiple Branch_____

The statement:

ON K2% **GOTO** 300,400,500

will cause an unconditional branch to line 300, 400 or 500 if the value of K2% is respectively 1, 2, or 3. Any BASIC expression may be used between **ON** and **GOTO**. If the expression lies outside the given range (1-3 in this case), the statement may be ignored or execution may stop.

The IF-THEN Statement_____

IF logical expression **THEN** statement

If the logical expression is true (not zero) then the statement is executed.

Alternate forms:

> **IF** logical expression **THEN** line number
> **IF** logical expression **GOTO** line number

If the logical expression is true then the program branches to the given line number.

The IF-THEN-ELSE Statement

> **IF** logical expression **THEN** statement 1 **ELSE** statement 2

If the logical expression is true then the statement following the **THEN** is executed; otherwise, the statement following the **ELSE** is executed.

ITERATIVE STATEMENTS

The FOR Loop

> **FOR** variable = first-value **TO** last-value
> . . .
> **NEXT** variable

Example:

> **FOR** I = 1 **TO** N
> . . .
> **NEXT** I

Alternate form:

> **FOR** variable = first-value **TO** last-value **STEP** value
> . . .
> **NEXT** variable

Example:

> **FOR** I = N **TO** 1 **STEP** −1
> . . .
> **NEXT** I

The WHILE Loop

WHILE expression
. . .
WEND

Example:

WHILE delta > tol
. . .
WEND

INPUT and OUTPUT

Input Data from Keyboard

INPUT variable or variable list
INPUT "string prompt"; variable or variable list
LINE INPUT "string prompt"; string variable (input entire line)

Input Data from Program

READ variable or variable list
. . .
DATA item or items

Output Data to Console

PRINT variable list
PRINT USING string format; variable list

SUBROUTINES

Unconditional Transfer

A subroutine is called with a statement such as:

GOSUB 4000

where the argument is the first line number of the subroutine. A **RETURN** statement at the end of the subroutine returns control to the calling program. One subroutine may call another subroutine. However, since BASIC variables are global, a subroutine may not recursively call itself.

Conditional Transfer

Any one of a number of subroutines can be selected based on the value of an index. The expression:

ON K2% **GOSUB** 3000, 4000, 5000

will call the subroutine at line 3000, 4000, or 5000 if the value of K2% is respectively 1, 2, or 3. Any BASIC expression may be used between **ON** and **GOSUB**. If the expression lies outside the given range (1-3 in this case), the statement may be ignored or execution may stop.

Bibliography

Daniel, Cuthbert; Wood, Fred; and Gorman, John. *Fitting Equations to Data*. New York: John Wiley & Sons, 1971.

Fike, C.T. *Computer Evaluation of Mathematical Functions*. Englewood Cliffs, N.J.: Prentice-Hall, 1968.

Forsythe, George; Malcolm, Michael; and Moler, Cleve. *Computer Methods for Mathematical Computations*. Englewood Cliffs, N.J.: Prentice-Hall, 1977.

Fox, L., and Mayers, D.F. *Computing Methods for Scientists and Engineers*. New York: Oxford University Press, 1968.

Gilder, Jules. *BASIC Computer Programs in Science and Engineering*. Rochelle Park, N.J.: Hayden, 1980.

Grogono, Peter. *Programming in Pascal*. Revised Edition. Reading, Mass.: Addison-Wesley, 1980.

Hart, John F., et al. *Computer Approximations*. New York: John Wiley & Sons, 1968.

Hastings, Cecil, Jr. *Approximations for Digital Computers*. Princeton, N.J.: University Press, 1955.

Hewlett-Packard. *HP-25 Applications Programs*. Cupertino, Calif., 1975.

Hornbeck, Robert. *Numerical Methods*. New York: Quantum Publishers, 1975.

Hubin, Wilbert. *BASIC Programming for Scientists and Engineers*. Englewood Cliffs, N.J.: Prentice-Hall, 1978.

International Business Machines Corporation. *System/360 Scientific Subroutine Package, Programmer's Manual*. White Plains, N.Y., 1966.

Jennings, Alan. *Matrix Computation for Engineers and Scientists*. New York: John Wiley & Sons, 1977.

Kernighan, Brian W., and Plauger, P.J. *Software Tools*. Reading, Mass.: Addison-Wesley, 1976.

Khabaza, I.M. *Numerical Analysis*. Elmsford, N.Y.: Pergamon Press, 1965.

Kreyszig, Erwin. *Advanced Engineering Mathematics*. New York: John Wiley & Sons, 1967.

Ley, B. James. *Computer Aided Analysis and Design for Electrical Engineers*. New York: Holt, Rinehart and Winston, 1970.

McCormick, John, and Salvadori, Mario. *Numerical Methods in FORTRAN*. Englewood Cliffs, N.J.: Prentice-Hall, 1964.

Nagin, Paul, and Ledgard, Henry *BASIC with Style: Programming Proverbs*. Rochelle Park, N.J.: Hayden, 1978.

Ruckdeshel, F.R. *BASIC Scientific Subroutines*. Volume 1. Peterborough, N.H.: Byte/McGraw-Hill, 1981.

Scheaffer, R.L., and Mendenhall, W. *Introduction to Probability: Theory and Applications*. N. Scituate, Mass.: Duxbury Press, 1975.

Smith, J. *Advanced Analysis with the Sharp 5100 Scientific Calculator*. New York: John Wiley & Sons, 1979.

Sokolnikoff, Ivan, and Sokolnikoff, Elizabeth. *Higher Mathematics for Engineers and Physicists*. New York: McGraw-Hill, 1941.

Vandergraft, J.S. *Introduction to Numerical Computations*. New York: Academic Press, 1978.

Walpole, Ronald, and Myers, Raymond. *Probability and Statistics for Engineers and Scientists*. New York: Macmillan Co., 1972.

Wylie, Clarence, Jr. *Advanced Engineering Mathematics*. Second Edition. New York: McGraw-Hill, 1960.

Index

The SYBEX Library

BASIC PROGRAMS FOR SCIENTISTS AND ENGINEERS
by Alan R. Miller   340 pp., 120 illustr., Ref. B240
This second book in the "Programs for Scientists and Engineers" series provides a library of problem solving programs while developing proficiency in BASIC.

INSIDE BASIC GAMES
by Richard Mateosian   350 pp., 240 illustr., Ref. B245
Teaches interactive BASIC programming through games. Games are written in Microsoft BASIC and can run on the TRS-80, APPLE II and PET/CBM.

FIFTY BASIC EXERCISES
by J.P. Lamoitier   240 pp., 195 illustr., Ref. B250
Teaches BASIC by actual practice using graduated exercises drawn from everyday applications. All programs written in Microsoft BASIC.

YOUR FIRST COMPUTER
by Rodnay Zaks   260 pp., 150 illustr., Ref. C200A
The most popular introduction to small computers and their peripherals: what they do and how to buy one.

DON'T (or How to Care for Your Computer)
by Rodnay Zaks   220 pp., 100 illustr., Ref. C400
The correct way to handle and care for all elements of a computer system including what to do when something doesn't work.

INTRODUCTION TO WORD PROCESSING
by Hal Glatzer   200 pp., 70 illlustr., Ref. W101
Explains in plain language what a word processor can do, how it improves productivity, how to use a word processor and how to buy one wisely.

INTRODUCTION TO WORDSTAR
by Arthur Naiman   200 pp., 30 illustr., Ref. W110
Makes it easy to learn how to use WordStar, a powerful word processing program for personal computers.

FROM CHIPS TO SYSTEMS: AN INTRODUCTION TO MICROPROCESSORS
by Rodnay Zaks   560 pp., 255 illustr., Ref. C201A
A simple and comprehensive introduction to microprocessors from both a hardware and software standpoint: what they are, how they operate, how to assemble them into a complete system.

MICROPROCESSOR INTERFACING TECHNIQUES
by Rodnay Zaks and Austin Lesea   460 pp., 400 illustr., Ref. C207
Complete hardware and software interconnect techniques including D to A conversion, peripherals, standard buses and troubleshooting.

PROGRAMMING THE 6502
by **Rodnay Zaks**   390 pp., 160 illustr., Ref. C202
Assembly language programming for the 6502, from basic concepts to advanced data structures.

6502 APPLICATIONS BOOK
by **Rodnay Zaks**   280 pp., 205 illustr., Ref. D302
Real life application techniques: the input/output book for the 6502.

6502 GAMES
by **Rodnay Zaks**   300 pp., 140 illustr., Ref. G402
Third in the 6502 series. Teaches more advanced programming techniques, using games as a framework for learning.

PROGRAMMING THE Z80
by **Rodnay Zaks**   620 pp., 200 illustr., Ref. C280
A complete course in programming the Z80 microprocessor and a thorough introduction to assembly language.

PROGRAMMING THE Z8000
by **Richard Mateosian**   300 pp., 125 illustr., Ref. C281
How to program the Z8000 16-bit microprocessor. Includes a description of the architecture and function of the Z8000 and its family of support chips.

THE CP/M HANDBOOK (with MP/M)
by **Rodnay Zaks**   330 pp., 100 illustr., Ref. C300
An indispensable reference and guide to CP/M — the most widely used operating system for small computers.

INTRODUCTION TO PASCAL (Including UCSD PASCAL)
by **Rodnay Zaks**   420 pp., 130 illustr., Ref. P310
A step-by-step introduction for anyone wanting to learn the Pascal language. Describes UCSD and Standard Pascals. No technical background is assumed.

THE PASCAL HANDBOOK
by **Jacques Tiberghien**   490 pp., 350 illustr., Ref. P320
A dictionary of the Pascal language, defining every reserved word, operator, procedure and function found in all major versions of Pascal.

PASCAL PROGRAMS FOR SCIENTISTS AND ENGINEERS
by **Alan Miller**   400 pp., 80 illustr., Ref. P340
A comprehensive collection of frequently used algorithms for scientific and technical applications, programmed in Pascal. Includes such programs as curve-fitting, integrals and statistical techniques.

50 PASCAL PROGRAMS
by **Rudolph Langer and Rodnay Zaks**   275 pp., 90 illustr., Ref. P350
A collection of 50 Pascal programs ranging from mathematics to business and games programs. Explains programming techniques and provides actual practice.

APPLE PASCAL GAMES
by **Douglas Hergert and Joseph T. Kalash**   380 pp., 40 illustr., Ref. P360
A collection of the most popular computer games in Pascal challenging the reader not only to play but to investigate how games are implemented on the computer.

INTRODUCTION TO UCSD PASCAL SYSTEMS
by Charles T. Grant and Jon Butah   300 pp., 110 illustr., Ref. P370
A simple, clear introduction to the UCSD Pascal Operating System for beginners through experienced programmers.

INTERNATIONAL MICROCOMPUTER DICTIONARY
140 pp., Ref. X2
All the definitions and acronyms of microcomputer jargon defined in a handy pocket-size edition. Includes translations of the most popular terms into ten languages.

MICROPROGRAMMED APL IMPLEMENTATION
by Rodnay Zaks   350 pp., Ref. Z10
An expert-level text presenting the complete conceptual analysis and design of an APL interpreter, and actual listings of the microcode.

SELF STUDY COURSES

Recorded live at seminars given by recognized professionals in the microprocessor field.

INTRODUCTORY SHORT COURSES:
Each includes two cassettes plus special coordinated workbook (2½ hours).

S10—INTRODUCTION TO PERSONAL AND BUSINESS COMPUTING
A comprehensive introduction to small computer systems for those planning to use or buy one, including peripherals and pitfalls.

S1—INTRODUCTION TO MICROPROCESSORS
How microprocessors work, including basic concepts, applications, advantages and disadvantages.

S2—PROGRAMMING MICROPROCESSORS
The companion to S1. How to program any standard microprocessor, and how it operates internally. Requires a basic understanding of microprocessors.

S3—DESIGNING A MICROPROCESSOR SYSTEM
Learn how to interconnect a complete system, wire by wire. Techniques discussed are applicable to all standard microprocessors.

INTRODUCTORY COMPREHENSIVE COURSES:
Each includes a 300-500 page seminar book and seven or eight C90 cassettes.

SB1—MICROPROCESSORS
This seminar teaches all aspects of microprocessors: from the operation of an MPU to the complete interconnect of a system. The basic hardware course (12 hours).

SB2—MICROPROCESSOR PROGRAMMING
The basic software course: step by step through all the important aspects of microcomputer programming (10 hours).

ADVANCED COURSES:
Each includes a 300-500 page workbook and three or four C90 cassettes.

SB3—SEVERE ENVIRONMENT/MILITARY MICROPROCESSOR SYSTEMS
Complete discussion of constraints, techniques and systems for severe environment applications, including Hughes, Raytheon, Actron and other militarized systems (6 hours).

SB5—BIT-SLICE
Learn how to build a complete system with bit slices. Also examines innovative applications of bit slice techniques (6 hours).

SB6—INDUSTRIAL MICROPROCESSOR SYSTEMS
Seminar examines actual industrial hardware and software techniques, components, programs and cost (4½ hours).

SB7—MICROPROCESSOR INTERFACING
Explains how to assemble, interface and interconnect a system (6 hours).

SOFTWARE

BAS 65™ CROSS-ASSEMBLER IN BASIC
8" diskette, Ref. BAS 65
A complete assembler for the 6502, written in standard Microsoft BASIC under CP/M®.

8080 SIMULATORS
Turns any 6502 into an 8080. Two versions are available for APPLE II.

APPLE II cassette, Ref. S6580-APL(T)
APPLE II diskette, Ref. S6580-APL(D)

FOR A COMPLETE CATALOG
OF OUR PUBLICATIONS

U.S.A.
2344 Sixth Street
Berkeley,
California 94710
Tel: (415) 848-8233
Telex: 336311

SYBEX-EUROPE
4 Place Felix Eboué
75583 Paris Cedex 12
Tel: 1/347-30-20
Telex: 211801

SYBEX-VERLAG
Heyestr. 22
4000 Düsseldorf 12
West Germany
Tel: (0211) 287066
Telex: 08 588 163

notes

notes